PENGUIN BOOKS

ON THE HOOF

PENGUIN BOOKS
Published by the Penguin Group
Penguin Group (NZ), 67 Apollo Drive, Rosedale, Auckland 0632,
New Zealand (a division of Penguin New Zealand Pty Ltd)
Penguin Group (USA) Inc., 375 Hudson Street, New York,
New York 10014, USA
Penguin Group (Canada), 90 Eglinton Avenue East, Suite 700,
Toronto, Ontario, M4P 2Y3, Canada (a division of Penguin Canada
Books Inc.)
Penguin Books Ltd, 80 Strand, London, WC2R 0RL, England
Penguin Ireland, 25 St Stephen's Green, Dublin 2, Ireland (a division
of Penguin Books Ltd)
Penguin Group (Australia), 707 Collins Street, Melbourne,
Victoria 3008, Australia (a division of Penguin Australia Pty Ltd)
Penguin Books India Pvt Ltd, 11, Community Centre,
Panchsheel Park, New Delhi – 110 017, India
Penguin Books (South Africa) (Pty) Ltd, Block D, Rosebank Office
Park, 181 Jan Smuts Avenue, Parktown North, Gauteng 2193,
South Africa
Penguin (Beijing) Ltd, 7F, Tower B, Jiaming Center, 27 East Third
Ring Road North, Chaoyang District, Beijing 100020, China

Penguin Books Ltd, Registered Offices: 80 Strand, London,
WC2R 0RL, England

First published by Penguin Group (NZ), 2014
10 9 8 7 6 5 4 3 2

Copyright © Ruth Low, 2014

The right of Ruth Low to be identified as the author of this work in
terms of section 96 of the Copyright Act 1994 is hereby asserted.

Designed and typeset by Jenny Haslimeier, © Penguin Group (NZ)
Image page 4: South Canterbury Museum, #1805
Maps by Jan Kelly
Prepress by Image Centre Ltd
Printed in China by 1010 Printing

All rights reserved. Without limiting the rights under copyright
reserved above, no part of this publication may be reproduced,
stored in or introduced into a retrieval system, or transmitted, in
any form or by any means (electronic, mechanical, photocopying,
recording or otherwise), without the prior written permission of
both the copyright owner and the above publisher of this book.

ISBN 978-0-143-57151-3

A catalogue record for this book is available
from the National Library of New Zealand.

www.penguin.co.nz

ON THE HOOF

The untold story of drovers in New Zealand

Ruth Entwistle Low

PENGUIN BOOKS

*To my companions on the journey:
Mark, Charlotte and Kate*

CONTENTS

Foreword   6

Introduction   9

1. **The Age of the 'Golden Fleece':** Droving's heroic phase   12

2. **On the Move:** The drover's role in the emerging farming economy   36

3. **The Long and the Short of It:** Droving routes of New Zealand   60

4. **The Faces of Droving:** The drovers on the road   92

5. **The Craft of Droving:** Day to day on the road   120

6. **Trials A-plenty:** Problems on the road   150

7. **Partners on the Journey:** The drover's horses and dogs   176

8. **Tools of the Trade:** The necessities for survival on the road   198

9. **The Twilight Years:** The growth in trucking and the demise of droving   222

Epilogue   235

Contributors   239

Endnotes   249

Glossary   258

Bibliography   259

Acknowledgements   264

Index   266

FOREWORD

I am in awe of the book you are now holding and what it says about its author.

Ruth came into my life nigh on 10 years ago when she invited me to attend a meeting of stock agents in Feilding – she wanted to discuss the possibility of tracing the whereabouts of some old-time drovers who had worked in the area. The Manawatu was my old stomping ground: I had worked there as a farmhand straight out of school, before joining a stock and station firm called The New Zealand Loan and Mercantile Agency Company Limited (what a mouthful!) in the 1960s. I was intrigued by Ruth's proposal and agreed to take part.

Ruth outlined to the group that it was her hope and intention to record the life and times of the droving fraternity throughout the country, and the vital role they have played in New Zealand's history. A huge ask, I thought! But we made a start: I listened to the stock agents' stories, and was able to recount my youthful days at the Feilding saleyards.

The job of a stock and station agent sounds glamorous: calling on farmer clients, drinking copious cups of tea, eating lavish lunches and chatting up the farmers' daughters. While the reality was something quite different, that was the vision of every would-be agent. And so the stock companies employed stock clerks, mostly cocky young men with the inevitable 'gift of the gab'. It was their job to do the donkey work on sale days, which in Feilding were every Friday.

Sale days meant being picked up from home just as the sun was rising and spending the next 10 hours taking delivery of the stock. Some beasts were trucked in, but at that time – the early 1960s – most were brought in by drovers. I recall them all vividly: Charlie Sundgren, Dan Cantwell, Alby Burney, to name just a few. We had a love–hate relationship with these characters. We needed them and they needed us. We needed stock to be brought in from holding paddocks before the sale started, and we often needed them taken to the railhead once the sale was over.

If it's true that we worked long hours, the drovers worked even longer. Inevitably there was friction brought on by mutual tiredness, smart-aleck remarks, and arguments over the exact number of sheep or cattle in a mob. Nearly every time, the drovers had it right!

I left the stock firm in 1965 and joined the staff of the television show *Country Calendar* as a reporter, eventually taking over the running of the series. Partly due to nostalgia, perhaps, but mostly because they were such characters, drovers and droves featured prominently in our programme. But a television show, by definition, can only provide a superficial insight into the life of a drover. To go deeper takes a lot of painstaking research – and that's one of the reasons I'm in awe of what Ruth has achieved with this book. The hours of interviewing and reading have manifested in a comprehensive history of drovers and droving in New Zealand.

It takes more than just research, too. Even at that first meeting, Ruth impressed me with her uncanny ability to get people to tell their stories, to immerse herself in the tales and recount them in an authentic way. And, of course, time was against her. So many of the old drovers are now well into their twilight years. Some have been lost to us, but through talking to close family members, Ruth has been able to produce vivid accounts of their droving careers.

This is a book that defines the hardships, the frustrations and, at the same time, the satisfaction these drovers felt from getting the job done. Unsung heroes? Not any more – thanks to Ruth's dedication, her attention to detail and, above all, her skill in producing a thoroughly absorbing read.

Frank Torley ONZM
Former Executive Producer, *Country Calendar*

The Drover

Down in the cold grey town I know him,
A lean brown man with a keen blue stare;
He idles beside the dusty pavements
With nothing to do, and no one to care.
For he is dreaming of open country,
And long white roaming roads that go
By tussock plains with a deep-blue skyline
Where all the winds of Heaven blow.
Roadside camps, and the slow sheep huddled
Under rocks by the blazing noon,
Wisp of smoke from a dying campfire
Threading upward beneath the moon.
He is hearing the pattering murmur
Of patient flocks, and the creaking load
Of the pack-horse plodding beside his stirrup;
He smells the reek of the dusty road.
He may never return again to know them,
But he will remember until he dies
The long white roads of the open country,
And winds that blow from deep-blue skies.

—Joyce West (1939)[1]

INTRODUCTION

A drover, clothed in wool shirt, sleeves rolled up, tweed pants, heavy workboots, hat pulled down over his eyes to break the glare of the sun, fingers working on a 'rollie', reins from his mount loosely draped over a tanned forearm, ambles leisurely along the road. His horse follows docilely behind. A mob of sheep walks peaceably in front, nibbling at roadside grass as working dogs allow. An occasional shrill whistle or coarse verbal command redirects the dogs' attention.

Such a scene – albeit not always so idyllic – has been played out in New Zealand's farming history thousands of times. The romanticised image of man, dog and stock will be indelibly etched in the minds of many who have happened across such a sight while travelling in rural New Zealand. For some, perhaps, a less pleasant memory lingers of a mêlée, their vehicle engulfed in milling ruminants and the associated smells and noise, augmented by the verbal abuse from an irate drover. Idyllic or otherwise, such scenes are quickly becoming a distant memory as the role of the drover disappears from the rural landscape.

Droving – the movement of stock on the hoof from farmgate to saleyards, abattoirs, railheads, freezing works or another farm – is distinct from mustering. This history is not concerned with the many stockmen who worked the large stations with the regular mustering of livestock for weaning, shearing, marking and so on. The broader term 'stockman' embraces both roles, and many would have carried out both tasks, but the focus of this book is on the story of stock on the road and those who drove them.

Much has been made in farming history of the role of the shepherd or stockman working in isolation and facing hardship on stations throughout the country. The drover, on the other hand, is often barely mentioned. Yet their role was an important one when you consider the number of stock that were sold at saleyards annually – a number that increased exponentially from the time of settlement, and grew as the impact of the development of refrigerated transport took hold and freezing works sprang up around the country. Even with the advent of rail and, later, trucking, droving was still the main means of stock 'transportation' well into the twentieth century.

This books aims to redress that imbalance and to acknowledge the work of those men, women and children who drove thousands of head of valuable stock along the roads of our small nation. They were a significant cog in the huge agricultural mechanism that generated so much of our country's wealth. For the drovers there was no sense of anything extraordinary in what they were doing, yet for many there was a real love for their work. Despite the hardships – sleeping rough, meagre rations, cold and miserable weather, the run-ins with inconsiderate motorists, the loss of stock or dogs – if the opportunity arose to drove again, many would jump at it.

This book aims to introduce the reader to the drovers – from those who drove the very first sheep in to the pastoral runs to those who are the last of a dying breed – and to record the individuals on the road, their work, their tools of trade, the territory they covered and the challenges they faced. It also looks at the reasons for droving in the context of the emerging farming environment, and of droving's twilight years. Much of the material for this book is based on some 60 interviews with those involved in the droving industry. The interviews provide a colourful insight into what has become an almost forgotten field.

Within the drovers' ranks were the educated, and the uneducated, those who chose to go droving and those who 'fell' into it. There were shepherds, stockmen and the odd rogue or two. Women, too, feature in the history of droving: the long-suffering pioneer women who journeyed out with their men and a few head of stock to make a go of their allotted acres; the land girls in World War Two, 'manpowered' onto farms, many with no farming experience at all; and the girls who gave up aspirations of a career to stay and work on family farms, who drove stock to railheads or saleyards as part of their farm duties. In the 1970s a few women could be found out in front of mobs of cattle, driving a truck towing a caravan. There were also those who might have only been droving a few times and who would not classify themselves as a drover at all, but the memories of those experiences would always awaken a spark in them. The experiences offered a freedom from the routine of everyday life, an element of adventure with the rough and ready lifestyle, the unpredictability of working with stock, and perhaps a bit of romanticism – not that they would readily admit to that.

For some farmers, droving was a part of the rhythm of farming life: perhaps it was not economically viable for them to employ a drover, or they were unwilling to rely on others, so they drove their own stock to market.

For some, the title 'drover' was a badge of honour and a reflection of how they viewed themselves: it was not just what they did, but who they were. It spoke of the pride they had in ensuring the stock they took on the road arrived in better condition than when it left. It was about the horsemanship, the animal husbandry, the knowledge of the environment, their dogs. It reflected the love they had for their work and the good and bad it entailed. They may well have worked seasonally, mustering, farm labouring, shearing – but if you asked them, 'drover' would be the title they gave themselves.

Whether holding court at the local, having a quiet chat to a cockie as the stock quietly meandered past, or sitting with a mate around a fire at the end of a long day on the road, the drover was never short of a story or two. Their work lent itself to the telling of yarns, the amber liquid loosened their tongues and, while perhaps like a fisherman there was the odd exaggeration, opportunity to expound on their daily experiences was seldom missed. Yarning has been an intrinsic element particularly of our male culture since pioneering days, and stories abound in these pages from those who drove. Battles with motorists, the perils of river crossings or the challenges of pushing stock through towns were all grist to the mill; along with tales of their quick wit and acerbic tongue extracting them from all manner of dire circumstances. And of course no drover would be worth his weight without a story or two about the superhuman abilities of his dogs.

Ever present as the drovers were interviewed was the shadow of another time, a time past, when life was lived at a much slower pace. The speed of technological change has mirrored the increasing pace of our lives, and is at variance with the speed of the drover and his stock. Before the memories of that time are lost, it seems important here to allow the voices of the drovers to be heard.

Sheep drovers at Pakowhai in Hawke's Bay, circa 1920. *Alexander Turnbull Library, 1/1-007287-G*

1. THE AGE OF THE 'GOLDEN FLEECE'

Droving's heroic phase

Most of the settlers . . . would do wisely in the opinion of all old settlers here to immediately purchase stock, and merely make the 50 acre section a garden for family and domestic animals. The truth of this is pretty well proved by Stafford's case – he wasted no money in buying lots of land and paying high wages for labour in the first days of the settlement but bought sheep and quietly let them feed and increase till he had the means of cultivating his land, whereas most others squandered away their money in useless labour, and after all had nothing to carry on with . . .'[1]

As had been proven in the young colony of Australia, there was money to be made in growing wool. A non-perishable commodity with ready markets in the British and European textile industry, wool was a means to wealth. Those with an eye for profit in the new British colony of New Zealand saw the promise of wealth in the tussock country in the east of both islands. The large tracts of open country were suited to extensive pastoralism, and money could be spent on purchasing stock rather than immediate land purchase, or being squandered on labour necessary to break in often unprofitable small farms.

Thousands of sheep were driven onto newly leased runs from the time of the first pastoralists taking sheep into the Wairarapa in 1844 until the last runs were taken up in the South Island east of the Main Divide by 1865. Explorers and surveyors were still negotiating the deeper reaches of the country as would-be pastoralists began droving their mobs of merinos onto the tussocklands. The men who drove the stock into the eastern plains showed great courage and determination. With minimal supplies and little knowledge of the routes, they faced hardship and exposed themselves, at times, to considerable danger. Their efforts can be seen as heroic in light of the problems and dangers they overcame, and the financial benefit to the fledgling colony that the eventual establishment of pastoralism brought.

The small towns and communities that were being established were connected by little more than dirt tracks. Most communities were coastal and relied on sea travel for communication; inland travel was only for the hardy. Travellers negotiating inland routes ran the risk of losing their way among the maze of mountain passes and hills. Travelling with hundreds of sheep meant longer trips and greater discomfort. The fast-flowing rivers, prone to flooding, were a constant source of danger and frustration. Finding a safe place to cross and then convincing recalcitrant sheep to enter the water could add days to a drove.

An air of adventure permeates such trips nevertheless – with monotony an equal bedfellow.

> To cure an unfit new-comer, dangerously enamoured of the romance of colonization, few experiences could surpass a week of sheep-driving, where life became a prolonged crawl at the heels of a slow, dusty, greasy-smelling 'mob' straggling along at a maximum pace of two miles an hour . . .[2]

Three intrepid journeys described here represent the tenacity, determination and fearlessness of the drovers. The first is the drove of the 'quintet' into the Wairarapa in 1844; the second is that of Jollie and Lee from Nelson to the Canterbury Plains in 1852; and the third is the drove of Alfred Duncan with William Gilbert Rees's sheep to Lake Wakatipu and what later became Queenstown. The significance of the Wairarapa and Otago droves lay in their being the first droves into a region. The drove of Jollie and Lee had even greater significance: this route provided the pastoralists of the Wairau with an inland route to Canterbury, and thus a ready means to drive surplus sheep to the mushrooming market on the Canterbury Plains. For Cantabrians it provided a steady supply of sheep to stock quickly establishing runs.

It is important to note, when talking about the east coast grasslands, that there was no uniformity of vegetation in the pastoral regions. Even within regions there could be great variances in vegetation and soil fertility. Drovers would be passing through country dominated by any of a variety of plants: tussocks, fern, manuka, toetoe, flax, cabbage trees, and the formidable speargrass (or 'spaniard') and equally ferocious matagouri (or 'wild Irishman'). These last two plants often formed an impenetrable wall, dealt to by torching it or finding an alternate route. The coarse tussock did not, in fact, provide much sustenance for the sheep; but it provided shelter for finer grasses and herbs, anise, bluegrass and the like, which stock thrived on.

The sea of tussock offered drovers little in the way of landmarks to guide them, and there was always the risk of becoming disoriented. John McGregor recalls in his reminiscences, when working as a shepherd for Henry Ford of Grampian Hills Station in the Mackenzie Country, taking 'straight lines by cabbage trees for Ashburton' while droving southward.[3] Even as the country became more settled, other vegetation was still to trouble drovers: six-foot-high flaxes on either side of the road through the township of Temuka hindered McGregor's progress with his sheep.[4]

The first drove into the Wairarapa

The first pastoral push took place in the North Island. Frustration with the lack of suitable land around the newly settled Wellington led those with capital and a growing awareness that there was money to be made 'off a sheep's back' to search further afield for pasture for their sheep. By 1843, parties of would-be pastoralists were eyeing up the Wairarapa. Reconnaissance trips led them to believe that, despite the difficulties of the route, the Wairarapa held potential: 'Many most trustworthy gentlemen have reported on its merits. They all agree that it contains many hundreds of thousands of acres of fine fertile and well watered land; about three fourths of it free from forest.'[5] Dissatisfied with the slow rate of progress with land purchase through the New Zealand Company ('the Company') and Government, aspiring pastoralists initiated leases for land with Wairarapa Maori. A. G. Bagnall in his history of the Wairarapa names 'the founding quintet' of pastoralists in the region as Clifford, Vavasour, Petre, Weld and Bidwill.[6]

Having earlier reconnoitred the district and organised leases with local Maori, the immediate goal for the quintet was to stock their land. They were fully aware of the slow progress being made on the Company road over the Rimutakas – and thus the difficulties of the inland route – so they took the coastal route instead. Two parties with their sheep left one after the other. Charles Bidwill departed at the end of April 1844, with William Swainson Jnr and 'a man and boy'. They each carried 'necessaries' on their backs and quart pots and pannikins slung on a belt around their waists. Two pack mares carried blankets and provisions and, most importantly, a 'net to fold the sheep at night'. They set out for Kopungarara with a mob of 350 merinos. William Vavasour, Henry Petre and Frederick Weld, very much a 'new chum', along with a shepherd and a boy and some men to carry the provisions, set off for Wharekaka with a few hundred sheep that had come by sea from New South Wales. Both Swainson and Weld wrote of their trips. While their accounts are skeletal in places, they offer glimpses of the hardships and privations of these remarkable droves. The droves, which were breaking new ground for others to follow, covered in some nine days[7] what today might be considered a short distance – 50–60 miles (80–100 kilometres): considering the terrain, this was quite an achievement.

Voicing the excitement and enthusiasm of a new chum setting out on a novel adventure, Weld recounts:

> It was on the 1st of May, nine days after landing, that I began my experience of bush life. I started on that bright May morning full of joy and hope, and in the best of health and spirits. Wellington Bay was a glorious sight in those days. Thick forests clothed the hills, now

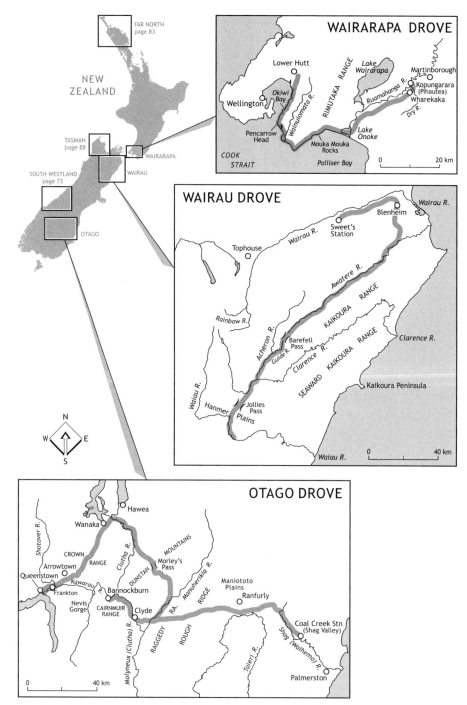

An approximation of the routes taken on the following historic droves. Top: The quintet's drove into the Wairarapa. Centre: Jollie and Lee's drove from the Wairau through to Canterbury. *Based on maps in A. D. McIntosh's* Marlborough: A Provincial History, *2nd edn, Capper Press, Christchurch, 1977, p. 120; and Martin Hill papers available from the Marlborough Museum Archives.* Bottom: Duncan's drove from North Otago through to Central Otago.

bare, in some places down to the very water's edge, and the snow on the summit of the Tarirua [Tararua] Hills made a perfect setting for the trees and the glittering foliage of the laurels and other evergreens in the foreground. Petre had put me up the night before we started at his house, so he and I set forth together, each with his roll of sleeping blanket, and a few indispensable articles wrapt up in it, on his back. We were joined at the other side of the bay by Vavasour, and here we found the flock of sheep in charge of a shepherd and boy, and some men who had been hired to carry flour, salt pork, cooking utensils, guns, axes, and such things. The hills were uncommonly rough and steep, the sheep weak after their journey, and the men's loads heavy, so we only reached the lagoon beyond Pencarrow Head the first night, and there under the side of the hill I passed my first night of camping-out. It was cold and windy, but I woke well-refreshed . . .[8]

Swainson looks back on the trip, writing to Bidwill's son some forty years later:

As far as I recollect our first day's journey was as far as Okiwi Bay where Mr. Donald, afterwards of Manaia, then lived, and we camped on the beach close to a rocky point which we could only pass at low or half tide, which would be about break of day next morning. We made a good fire on the beach, had our supper, smoked our pipes, and after seeing the horses were safe on the tether, curled ourselves in our blankets and went to sleep . . .

After passing this rock the track followed the ridge of hills nearest the beach where Clifford's sheep were running, and then again on to the beach at the mouth of the Wainui-o-mata and from there followed the coast to the mouth of the Wairarapa Lake. The stages we made were but short, both on account of there being no track but that between high and low water, which was very stony the greatest part of the way, and also there being plenty of feed for the sheep we did not hurry them, indeed people as a rule did not hurry themselves or their stock in those days.[9]

The greatest difficulty that both parties encountered was the logistics of moving stock around the Mouka Mouka (Mukamuka) Rocks. Weld brushes aside their experience in a few sentences:

We were many days reaching Wairarapa Valley, which was our destination, and lost some sheep on the way on the rocks, and in getting them round the headlands which were washed by the sea, there being no road inland. We encamped there in a grassy gully, and here we met another flock of sheep, belonging to my future neighbour, Mr. Bidwill, in charge of a Mr. Swainson.[10]

Swainson's account, however, elaborates and gives a greater sense of the harsh environment they battled:

Well, we got along very well until we reached the Mouka Mouka Rocks, which were the great obstacle in our road. There were three points of rock jutting out into the sea, two of which could be passed dry-footed at low water, but the third was always washed by the sea (though Colonel Wakefield, the New Zealand Company's agent, could not sanction the squatters going to the Wairarapa, still he did not see why he should not make a hole in the rock which they might use if they liked, so shortly previous a party was employed to make it passable, but they did very little good), and could only be passed at low water, and then only by getting a ducking if a wave higher than its fellows caught you, indeed when blowing south-east it was impassable for days, and the hill at the back too precipitious [sic] to admit of being climbed over, and many an unlucky traveller had been detained there for days together. Subsequent earthquakes have, I understand, so raised the beach that this post is passable at any time. Arrived here we had to catch every sheep, and standing in the water pass them from hand to hand round the point, a work of time, but which was accompanied with less trouble than we expected, and without any loss.[11]

After such exertion Swainson writes of sleeping despite heavy rain 'so thoroughly tired was I'.

The harsh weather made life less agreeable:

. . . for I remember waking with a stream of water running under me, and your father sitting with his blanket over his head and saying to me in a very injured tone of voice, 'Well, I never did see such a beggar to sleep.'[12]

It also delayed the drove:

Next morning we found that the Lake had opened its passage into the sea, and that therefore there was no chance of getting across even in canoes, and the rush of water would be for some time too strong

to do so with safety. We therefore moved to the edge of the Lake, to
where I believe Matthew's property now is, and from there your father
returned to the Hutt for some few cattle.[13]

It was at this point that the Weld party caught up with Bidwill's, although
they were careful not to mix the flocks.

The drove was further delayed by a hitherto unknown experience: the
need to negotiate with local Maori to ferry the men and their sheep across
the lake. The significance of such negotiations for future trips was not lost
on Weld as he writes of waiting for the negotiations to be settled:

They [Bidwill's party] were travelling in the same direction as we
were, and, like us, were waiting till an arrangement could be made
with the Maories to ferry the sheep across the outlet of the lake into
the new district. At last the Maories, who are the keenest hands in
the world at a bargain, especially in a case like this when they knew
we were entirely dependent on their services, came to terms. This
was a bargain of much importance, as it was bound to fix the price of
such dealings in the future, and so seriously affect our access to our
market. Some days, therefore, were taken over it, and at the end we
had to give about twice as much as would have been asked by any
English ferryman.[14]

With the prices settled, Swainson waited for Bidwill to return from the Hutt
with the cattle while the Wharekaka party crossed the lake and were the
first to settle on their station. Swainson and Bidwill soon followed, though
there is no account of moving the cattle as well as the sheep. Swainson
remarks that while he did not remember each stage of the remainder of the
trip, he did recall that:

The difficulties were all about the same size and were thick, although
we had the benefit of Weld's preceding us. There was next to no native
track and in trying to avoid places where we could see Weld had been
in difficulties we got into worse, especially in finding crossing places
to the swampy gullies and creeks for the two pack mares, Weld having
had none, but more men to carry necessaries.[15]

With no road through to Kopungarara, Bidwill's party stayed at Wharekaka
until the 'natives' had cut a route through the bush. While the initial drove
was over, the privation, of course, did not end as they worked to develop
their new runs.

As the first two parties took up their land, others quickly followed. Within five years, Census figures show the Wellington region – which included the Wairarapa – as having over 35,000 sheep: an increase of nearly 30,000 from 1844.[16]

> A push into Hawke's Bay began as land became scarce in the Wairarapa. Frederick Tiffen travelled up with the first sheep, some 3000 head, to Pourere in January 1849 from his brother's run near what later became known as Carterton. With the growing understanding that there was money in sheep, there was growing pressure for large tracts of land to come under European control, and in the push further north, more of the eastern coast was taken up. By 1852, through stocking of new runs and natural increase, there were some 60,000 sheep in the Wellington region, including Hawke's Bay.[17]

The inland route to Canterbury

The tussock country of the Canterbury Plains naturally lent itself to pastoralism. Within the first year of its settlement in 1850, 20 runs were taken up, on around 292,000 acres. By 1853, at the time of the establishment of the Canterbury Provincial Council, nearly 1 million acres of land was taken up.[18]

A ready supply of sheep was needed to stock these runs. A drought in Australia forced Australian runholders to destock, and this helped to boost numbers. Closer to home the Wairau Plain, which was quickly settled when Governor Grey purchased it in 1847, also provided thousands of sheep. First, however, they needed to find a suitable stock route from the Wairau to the Canterbury Plains.

The mountains, valleys and rivers from the Wairau through to Canterbury confounded efforts to find an inland route. Edward Lee was the first to successfully navigate a suitable stock route at the end of 1851, when he was searching for land for a run. It was 1852 before the first sheep were driven through. Lee, accompanied by the surveyor Edward Jollie, successfully drove 1800 sheep from the Nelson district through to Canterbury. Lee was fully aware of what lay ahead of them: his previous reconnoitring had thrown up all manner of obstacles. As his journal entries – printed in the *Lyttelton Times* in January 1852 – record, the trip was not for the faint-hearted: 'I am truly glad that this rough country is all passed through, every step one takes is done in fear and trembling . . .'[19]

> The initial part of the route that Edward Lee followed was over Barefell Pass, which had been discovered by Frederick Weld, who was searching for a route to take stock from Flaxbourne Station to a new run, Stonyhurst, in North Canterbury. A coastal route between the runs was not deemed possible. Weld believed he had found the route through to Canterbury but had mistaken his actual position in the maze of rivers and tributaries: 700 sheep were lost as a result. The first sheep onto the Stonyhurst run – some 1500 ewes – were, in fact, taken down from Cape Campbell along the coast to below Kaikoura.[20]

Jollie, riding his faithful horse Charley, left from Nelson. He travelled for three days to pick up his 900 sheep at Sweet's station[21] in the Wairau, joining his mob to Lee's.

> 'Charley' was a fine old horse, very good in a river, sure on his legs . . . well made – but not handsome, and a good slave on a journey . . . He could carry a heavy pack, as I soon proved by placing upon him huge saddle-bags, which I filled with stores to be used on our overland journey, and between the bags I mounted, having first of all strapped my roll of blankets in front of the saddle, and a long tether rope round 'Charley's' neck. I fancy I was a queer sight as I rode through Nelson after getting hoisted into my saddle at the door of the 'Whakatu' Hotel, and certainly I did not feel either happy or comfortable . . .[22]

Lee and Jollie set out in March 1852 with 1800 sheep. With them were:

> . . . a shepherd named Simpson who was drowned about four months afterwards in the Clarence River, and a man named John Berry, who had been one of my men when surveying in Christchurch. We had also three sheep-dogs and three horses and a foal with us. Our stores were flour, tea and sugar as necessaries and as luxuries a cheese and a lot of fine onions which Lee abstracted from a friend's garden which we passed on the road.[23]

They were later joined by a Canterbury settler, Augustus Percival, who caught up with them: as Jollie recalls, 'though of very little use to us in getting the sheep along, he was a pleasant companion sometimes, but sadly incommoded us in a small tent which was at first made quite little enough for four persons'.[24]

With knowledge gleaned from Lee's previous adventure, the men set off on their march to the Canterbury Plains. The pattern for the trip was for Lee and Simpson to focus on the droving, while Berry cooked and packed up. Jollie assisted where necessary, but his most valuable assistance was no doubt his afternoon reconnaissance, seeking the best possible route for the following day: here his surveying skills would have been useful. Progress was slow, 10- to 12-hour days, sometimes covering as little as 3½–4 miles (5½–6½ kilometres) in a day; in the end it took over a month to reach the plains. This was little wonder, as much of the land had not been burned off and was thick with matagouri and speargrass. As Jollie recalls:

> Our chief trouble in driving was getting the sheep over the rivers until they got used to it, the 'wild Irishman' a prickly shrub and the Spaniard – a very strong spear grass – also delayed us very much and the sheep did not get used to them, but on the contrary the more they felt them the less they appeared to like them. These noxious plants also delayed us by making many of the sheep lame, so that we always had a lame lot at the rear of the flock, which neither the dogs nor men could move faster than they thought fit.[25]

The trip was by no means straightforward, as sections of Lee's original route proved unsuitable for sheep. Lee and Jollie later found an alternate route after droving the sheep over the mountains to the Clarence River; this led them to what became known as Jollies Pass, located just north of the Hanmer Plain.

> We followed the line which Lee had travelled until we got to where the 'Guide' – a stream running from Barefells Pass – falls into the Acheron River. Lee had gone down the Acheron but as he did not think we could get sheep down on account of the thick Wild Irishman scrub, we crossed the Acheron, and I went ahead ascending a range on the right, and going along it for about four miles, arrived at a point from whence I could see through an opening in the Snowy Hills, the yellow grassy hills beyond the Hanmer Plain. I was very pleased with this view, and took the compass bearing of the opening, the direction of which we in future drove the sheep as near as the nature of the country would allow. It was near here that we crossed the sheep into a valley which we christened the Spanish Main on account of the intense number of 'Spaniards'; and as it was otherwise very thick with snow grass etc., we decided to take the sheep over the mountains to the Clarence River, while we sent the horses in charge of Percival and Berry through the

The harshness of the early droving environment is epitomised here in a scene sketched by Laurence Kennaway on his 1854 drove from the Hurunui River through to mid-Canterbury. Despite driving rain, constant watch had to be kept over the mob of sheep to prevent them from scattering. *From Laurence J. Kennaway's* Crusts: A Settler's Fare Due South, *Capper Press, Christchurch, 1970*

Spanish Main to the Clarence, and down that River to a point where we expected to meet them. We drove the sheep all day over the mountains and towards evening descended to the Clarence just above the junction of that river with the Acheron.[26]

After a hard day they were keen to find Berry and Percival, who had set up camp. This was not to be, however. Having left their sheep they went in search of the horses 'so that we might get some supper and our blankets, but darkness coming on we could not find them. So we lighted a fire and passed a cold night on the mountain side with empty stomachs'.[27] The next day Jollie, sure of their location, was able to convince Lee that it was necessary to head up the Clarence rather than down. After a 'very prickly, painful walk of two hours' they met up with Berry, who was 'coming to look

Alfred Chapman's sketch of breakfast being cooked outside the tent on his drove from Hawke's Bay to Wellington in 1856. *From Alfred Chapman's Journal from Hawke's Bay to Wellington: 4 Jan – 18 Feb 1856, Collection of Hawke's Bay Museums Trust, Ruawharo Ta-u-rangi, 6642*

for us with refreshments in his hands'. When they reached the campsite the group realised the significance of their location: they had found a pass, 'a great gap in the mountains' through to the Hanmer Plain. They faced one last challenge before their 'difficulties were over'. Jollie and Percival reconnoitred the area.

> On the top of the Pass, the vegetation was composed of stunted forest trees, about two feet high, with horizontal branches ten or twenty feet long, but on descending to the Hanmer Plain, we got into a black birch forest, which, however, by keeping on a ride we soon passed through. After descending about 1500 feet, we reached the level of the Hanmer Plain, satisfied that with a Lucifer match, and a few cuts with a hatchet, we could get the horses and sheep through. After resting awhile we set the grass and scrub on fire, and then we ascended the Pass, and got to our camp well satisfied with our day's work.

> The next day the fire was still raging furiously, so we left the sheep where they were on the other side of the Clarence, which was a boundary the fire could not cross. It was not for several days that the fire had sufficiently burned itself out to allow us with safety to move the sheep. In the meantime we employed our leisure in clearing the road through the black birch bush and other parts of the Pass, so that we eventually had no great difficulty in landing the flock safely in the Hanmer Plain . . .[28]

The small group made the trip stoically despite the privations; and remarkably, they lost only 50 or 60 sheep.

Jollie and Lee's legacy was their discovery of a valuable route between the regions, which allowed for the overlanding of thousands of sheep to stock fledgling runs. The route was shortened by several days in 1855 when Frederick Weld found a connection between Tophouse at the head of the Wairau Valley and the Acheron track to Canterbury.[29]

While heroic efforts to establish runs were being made throughout Canterbury, one man became famous for his droving exploits for all the wrong reasons. There are more questions than answers in relation to James Mackenzie's activities, but it seems certain that the legendary Scot uplifted 500 sheep that went missing from the Rhodes brothers' Levels run at Timaru and drove them down to Otago. In 1855 his luck ran out when he lifted a second mob of 1000 sheep from the Rhodes' station. He was caught in the country that is now named after him, and taken to Lyttelton for trial. All sorts of legends have sprung up about Mackenzie and his dog, as A. E. Woodhouse points out in her history on the Rhodes of the Levels:

> One cannot deny however, that man and dog were masters of their craft. It was no mean feat of sheepmanship to drive 500 ewes from the Levels to South Otago, through the wild back country, swimming across the big rivers, nor to lift 1000 ewes from the Taiko Flat, drive them 40 or 50 miles over trackless hills and gullies in four days with no assistance until more than half the distance was covered and to handle them so well that Sidebottom [the station overseer] did not report the discovery of a single exhausted sheep left by the way.[30]

Stocking the Rees run at Lake Wakatipu

Another drove of epic proportions was recorded in some detail by Alfred Duncan in *The Wakatipians or Early Days in New Zealand*. Duncan arrived in Lyttelton in March 1860, and soon moved to Otago where he sought a cadetship with William Gilbert Rees. With one drove under his belt, Duncan's 'first experience of travelling with stock . . . was of a sufficiently rough nature to give me a good idea of what lay before me in the life which I had chosen'.[31] No longer a cadet but an employee,[32] fired by 'Mr Ree's description of the lake scenery, and the wild life which would be the portion of those who first squatted there . . .'[33] he readied himself to assist on a drove from Coal Creek Station in Shag Valley to a site on Lake Wakatipu that later became Queenstown.

The Rees trip began in December 1860. The party included seven men, 13 horses and 3000 sheep.[34] Simon Harvey headed the party, Archie Cameron, 'a big Scotchman', helped with the driving, Andrew Low and Duncan were 'principally in charge of the sheep', George Simpson was the cook and packer as well as being in charge of the horses, and John Gilbert (William's brother-in-law) went ahead of the party to give notice of stock coming through. Harry Burr also came, but as to his part, Duncan was unsure: 'I can scarce define, as he accompanied us more for amusement than anything else, and left us, to go to South America, before we reached the Lake'.[35]

They travelled from Coal Creek over the Maniototo Plains, across the Rough Ridge and 'Ragetty' (Raggedy) Range, swimming the sheep across the Molyneux River[36] to Clyde. Duncan makes little of this early part of the drove. It may have been relatively uneventful, but the terrain would have been difficult and there were no roads, only 'a few tenuous tracks'.

It was when they arrived at what later became known as the Bannockburn that their troubles started. The sheep were rested while the men searched for a possible route along the Kawarau River. When Duncan was forced to head towards the snowline after being stopped by the 'steep gorge of the Nevis Creek', he found the natural rock bridge of the Kawarau: the overhang of the rocks enabled a person to jump across the gap to the other side of the river – that is, if they could face 'the turbulent water that lashed itself into foam on the rugged rocks below'. The men took turns scouting for a crossing, with a lot of effort and little gain – and little comfort awaiting them when they returned after a day of searching.

> It was nearly one o'clock in the morning when, tired and hungry, I reached the camp, where I found George on watch, and, having partaken of a pannikan of tea, some cold meat and damper, I crawled into the tent and was soon fast asleep between the folds of my blanket.[37]

William Rees and his partner, Colonel Grant, arrived at Duncan's camp a week later. Rees quickly took charge, and decided to cross the Kawarau a mile above its junction with the Molyneux. This decision nearly cost Duncan his life; and after the near-drowning they were forced to retrace their steps over the Cairnmuir Range.

> With this object in view we collected all the dried coorraddies (flower stalks of the flax bush) that we could find, and after tying them into bundles, lashed these together in the form of a raft, on which to take the rams, the stores, and the non-swimmers across. We were well provided with tether ropes, and these were brought into use, wherewith to pull the raft backwards and forwards across the river. Everything being ready, Mr Rees, who was, and I daresay is to this day, a more powerful swimmer, stripped off his clothes, and tying one end of the rope round his waist plunged into the water and struck out for the opposite side. The current is very strong here, but, in spite of that, he succeeded in making a tolerably straight passage across, and made fast the rope to a stout shrub on the bank. We then launched the raft, with nothing on it, on a trial trip, but no sooner did it get into the current than it was swept down, and we all, some six or seven, held on like grim death, in the hope of saving our valuable craft from the remorseless torrent. In vain we struggled, inch by inch the rope slipped through our fingers; Harry Burr, with his clothes off, stood by ready to plunge in to do something or other if he could only determine what, when, without a word of warning, all hands let go their hold with the exception of myself, and, as I was front man, and standing in the water, the sudden tightening of the rope lifted me right off my feet, and plunged me into the river about twenty feet from the land, where the rope held me under, it being tied fast on the bank, and thus the events of the next hour or so must be recorded by me from information which I afterwards received. The moment I disappeared Harry dived into the water and struck out for where I was last seen; grasping me by the clothes, he strove to drag me from under the rope, but instead of succeeding in doing so, the current sucked him under it likewise, and he also disappeared; at the same time Mr. Rees, from the opposite bank, shouted 'Cut the rope, cut the rope'. This advice was promptly acted upon, and, as the knife was drawn across the hemp, the raft shot away and Harry rose to the surface, with a good firm hold of my collar, and struck out for shore, where he was soon relieved of his burden, and I was spread out on the bank to disgorge the water of the Kawarau that I had swallowed, and to come to my senses again.

I was none the worse of my ducking, and strange to say, my watch was none the worse either . . .[38]

On Christmas Day 1860 they successfully crossed the sheep back over the Molyneux, standing in the river for up to 16 hours, working from dawn to dusk. Duncan reflects on the strange way in which he spent his Christmas, snatching only a few minutes to have a bite of food and to drink a pannikin of tea. The days were long and hard, and sometimes gruesome; ewes that were supposedly not in lamb started lambing, and Low and Duncan killed some 300 newborns; 'of course, it was quite impossible for us to take those on the toilsome journey which lay before us'.[39]

After passing the 'Shenan's' (Shennan) Station at Manuherikia and Thomson's run they crossed the Dunstan Mountains over Morley's Pass (later Thomson's Gorge); they slept at the pass on the only level and spongy place to pitch a tent in what turned out to be two inches of water, and dejectedly left the next morning without breakfast and with only a few scones in the pocket, as the water for tea had frozen to a solid block of ice in the billy. From there up to Hawea, crossing the Molyneux again, and along the Arrow and Frankton flats, which Rees had burned off on his first reconnaissance trip,[40] and then on to the 'promised land' over the Crown Range. All trials and hardships were forgotten, however, as they reached the top of the Crown Range and Duncan saw, for the first time, 'the Promised Land of Rees':

> . . . but, when we did at last reach the top of the range, a sight burst upon our view which caused us, one and all, to utter exclamations of pleasure. Nearly three decades have gone by since I stood on the summit of the crown range, and looked for the first time on the Promised Land of Rees, and yet I have only to close my eyes in order

---

The inner reaches of Otago district were explored in the early 1850s, but it was not until 1856 that there was a push for runs in the interior. At the time when Central Otago was being surveyed by Chief Surveyor John Turnbull Thomson in 1857, the first runholders were facing the arduous task of setting up their runs. By 1858 all the runs were claimed. As Grahame Sydney puts it in *Promised Land*, 'At the end of the decade, the once-silent valleys and ranges of the Interior murmured with the insistent bleating of many thousands of sheep, and the tents and first cob-and-thatch huts of leaseholders and their shepherds were widely sprinkled across the treeless, tussocky landscape.'[41]

to see the whole view start up before me as I then saw it. Away in the
distance the middle arm of the Wakatipu Lake lay glistening between
those precipitous ranges which, in these later days, have given it such
a character for grandeur of scenery. Nearer hand the Shotover and
the Arrow rivers flowed like silver threads through the blackened,
tomatagorra scrub-clothed plains which form the rich alluvial flats
lying between that part of the Lake, where Franktown now stands and
the gorge where Arrowtown now is . . .

As we stood gazing on the new land, not a sound broke the stillness
except an occasional gurgling bleat from some old ewe, as, with her
mouth full of aniseed plant, she intimated to a neighbour that she
approved of the pasture . . .[42]

The remainder of the drove went smoothly – even crossing the Arrow and Shotover rivers offered little trouble, as by then the sheep were like 'retriever dogs' with all the river crossings they had experienced. Once they were on the Shotover river flat, the journey was all but over.

The odyssey lasted close to three months and covered 200 miles (320 kilometres) of rugged terrain; and while it did not open up a significant stock route, it was remarkable for the sheer tenacity and perseverance shown by the drovers. In the case of Duncan, one cannot help but think that such a trip was a remarkable apprenticeship for a new chum.

The trials of droving

While these accounts give insight into the heroic efforts required to establish pastoralism within the colony, other scattered accounts also help create a greater understanding of the hazards faced by those who were moving stock on the hoof throughout this period. The nature of colonial sheep was cause for great consternation:

A word or two here about sheep in the colonial meaning of the word.
Colonial sheep are not English sheep. They are – though tolerably
well-bred – half-wild, unmanageable animals, accustomed to roam
over great tracts of open country, allowing no one to approach them
within half a mile, but stretching away at full gallop, and trying the
mettle of good dogs to head them.[43]

They needed to keep constant watch over the animals: it was easy for stragglers to get away or be left behind. Even at night there was little rest – with no such thing as fences, the drovers sometimes had to work in shifts through the night to prevent hungry stock wandering off. In fact, the sheep were so wild and restless that:

> . . . it was necessary to keep a double watch: two of us took charge at seven o'clock, the other two camping; – it being arranged that each pair should call the other after four hours' watch. This watching consists of continual walking, backwards and forwards, round the flock, from one point to another, to check the constant efforts of the sheep to break away. As soon as the watcher has stumbled over the rough ground to one end of the flock, an unquiet and suspicious bleat or two, and a rattle of tramping feet will summon him to the other, where a throng of grey nodding heads will be indistinctly seen, stringing out from the main circle, and trying to break away through the darkness. This attempt having been frustrated, a similar one calls him – like a human pendulum – to the other end, and then back again, and back again, and so on through the night; till the time at last comes for calling out the other human pendulum, who meanwhile has been enjoying his night's rest . . .'[44]

To such wearisome routines could be added inclement weather. For those sleeping rough it was just another discomfort to be faced – wet clothes, wet bedding, and impossible to light a fire. Drovers on the open expanse of the Canterbury Plains had the additional challenge of the hot, dry nor'westers that unsettled stock and made work tiresome – especially along a riverbed: 'During a nor'wester, the sand on the river-bed is blinding, filling eyes, nose, and ears, and stinging sharply every exposed part.'[45]

The hot and dry atmosphere of the nor'wester was replaced by the sou'wester:

> The force of the south-west wind is here broken by the front ranges, and on these it often leaves its rain or snow . . . On the plains, it will often blow for forty-eight hours, accompanied by torrents of pelting pitiless rain, and is sometimes so violent, that there is hardly a possibility of making headway against it.[46]

In such storms, when the sou'wester 'came tearing across the plains', the stock were likely to break away and scatter, and night vigils were necessary. Days could be added to a drove reassembling a mob that, once it had scattered, might only be stopped by a river or other obstacle.

The greatest threat to all drovers throughout this period, however, was river crossings. Many lives were lost in New Zealand rivers. They were

Right: The rugged terrain that Lee and Jollie travelled with their sheep eventually led them through what became known as Jollies Pass and onto the Hanmer Plain. Watercolour by Theodore Octavius Hurt, circa 1861–71. *Alexander Turnbull Library, E-501-f-015*

Below: After an epic journey, Alfred Duncan finally arrived at 'the Promised Land of Rees'. While it took some years to establish the station fully, reaching this site must have brought a great sense of relief. Watercolour by William Fox, 1864. *Alexander Turnbull Library, WC-327*

Seeing the potential in the Wairarapa, the 'quintet' negotiated leases with local Maori, and drove their sheep into new pastures. Lithograph by Samuel Charles Brees, circa 1843, showing Wairarapa Lake on the left and the Rimutaka and Tararua ranges in the background. *Alexander Turnbull Library, PUBL-0011-08-1*

Out of barren landscapes, towns and cities emerged. The developing road networks aided the drover in his work. Watercolour by Christopher Aubrey, circa 1896, showing an Inglewood street scene with Mount Taranaki in the background. *Puke Ariki, New Plymouth, A96.982*

Battling through spaniard (speargrass) and wild Irishman (matagouri) was an onerous task for those droving sheep through largely unexplored country. Burning it off was the generally favoured course of action. *Phil Bendle/nzwide*

A 1930s tourism poster depicting the South Westland drove captures the beauty of the terrain, while attesting to the significance of this drove in our history. Art print by Marcus King, circa 1930–39. *Alexander Turnbull Library, Eph-E-TOURISM-1930s-01; reproduced with permission of Tourism New Zealand*

Charlie Eggeling leads packhorse Oklahoma as they head towards the challenging and potentially dangerous Chasm Creek, near the top of the Maori Saddle. *Captain Maxwell Dowell QSM*

dangerous to cross at the best of times, but with the added complexity of crossing sheep, the risks increased considerably. Stories abound of days spent attempting to drove large mobs across the rivers; and the southern rivers – particularly the braided Rakaia and Rangitata rivers, snow-fed and unpredictable – were worst.

> These Mountain Streams are the great impediment and only varying feature of these plains, and the Rakaia is the worst of them. It is half river, half mountain torrent; nine tenths is shingle bed, with small irregular brooks, pursuing all kinds of irregular meandering; the whole space of river bed from half a mile to a mile in width – wild looking unmanageable waste, too wide to bridge, the streams too shallow and too rapid for a regular ferry, always changing their course, and so affording no place for fixed Engineering work; and in heavy freshes sending down such a body of water as to be perfectly impassable . . .[47]

The Rhodes family's huge Levels Station near Timaru nearly came to naught after their first futile attempts to cross their sheep over the Rakaia River. A mob of 5000 of their Banks Peninsula sheep being driven south in 1851 to stock their newly leased run refused to enter the water. After three days of failure, George Rhodes suggested to his brother Robert that they should turn back. Robert declared they would have 'one more try to-morrow'. The next morning they forced a few of the leaders into the water; the contrary animals followed and all crossed, then pushed onwards across the plains to their new home, becoming the first sheep in South Canterbury.[48]

Drovers had to resort to desperate means on occasion to get their obstinate charges across rivers. Laurence Kennaway in *Crusts,* his account of early settlement in New Zealand, talks of the droving experiences of himself and others. He recounts how a friend of his spent three days trying to cross 300 sheep across the Rangitata; at their wits' end, the shepherd and Laurence's friend carried them in pairs on horseback, 'fording the river time after time, till their horses were cramped and stiff with cold'.[49] The hard wet work, extra nights camping out, and the frustration of working with obstinate animals nearly reduced Laurence's friend 'to a state of raving idiotcy',[50] and it took him some time to recover.

Danger to sheep

It was crucial to the wellbeing of the sheep for the drover to keep a watchful eye on them. The drovers needed to be alert to many dangers, including: travelling at a reasonable speed so as not to overtire stock or increase the risk of injury; vegetation that might poison sheep (more of that later); wild dogs; and avoiding other stock in case they were contaminated with the debilitating disease, scab.

Scab, a disease in sheep caused by a mite burrowing under the skin, led to skin irritation and the eventual loss of wool and condition. The disease was first discovered in Nelson in 1845 from infected merino sheep imported from Australia. It quickly spread throughout the colony. It had the potential to spread wherever sheep were found en masse: with no fences separating runs, infected sheep could easily mix with clean sheep. Droving clean mobs through infected country also led to its spread.

> Early drovers had to watch out for wild dogs that worried their stock. Even as late as the 1920s wild dogs were a problem. Dick Gibbens, taking stock from the West Coast through to Canterbury before the Otira rail tunnel opened, was known to travel with a pistol to protect sheep at night against wild dogs; he slept under the bridge at Pegleg Creek with his pistol at the ready.[51]

The disease created a huge problem throughout the fast-developing sheep regions of the new colony. The Government made a determined effort to eradicate the disease by passing the Scab Ordinance of 1849, putting in place harsh controls by provincial governments and consolidating all legislation pertaining to scab in the 1878 Sheep Act. Sheep inspectors were appointed to prevent contaminated sheep from crossing provincial borders; sheep dipping was carried out annually; and drovers were required to warn all station owners before crossing their land. It helped, too, that more properties were fenced. Such measures ensured the eradication of the disease by the early 1890s.[52]

Some drovers took inordinate measures to avoid exposing their stock to infection, and this sometimes led to a completely different set of problems.

> On starting from the station from which he took them, my said brother drove by a circuitous route still further into the interior; for, it being impossible to get the sheep through in any other way, without passing infected country, he had determined, – backed up by my staunch old friend C., – to push his way over a rugged branch range of mountains which had never yet been crossed with stock. On arriving, at length, at the base of the lower spurs, he and his party were brought

to a difficult and unexpected halt; for the country was so tangled and overgrown with rough growth and scrub, that – as it was – it was impossible to force the sheep or even the horses through, by any possible means. Going back was not to be thought of, with scab and failure in the rear. C. and he resolved, therefore, to fire the jungle, and find afterwards their way across, through the burnt country, at the risk of being without feed for their stock on the route.

Toward evening, they drew the sheep down into the bed of a small river which wound under the base of the range; and leaving the two hands they had with them to keep the flock closely rounded up, they broke down low, withered shrubs with bushy heads, and lighting them into a blaze, drew them along the ground at the edge of the rank growth which clothed the hills. The wind was favourable, and by lighting at intervals for about a couple of miles, the whole line caught, and a hot, roaring belt of fire blazed up the hill-side, hissing and crackling, – and as C. declared, absolutely screaming human screams, – as the thick untouched jungle of scores of years withered and twisted in the flames.

All night long the fire roared way upon the hill-side, and all night long their solitary camp in the river-bed was lit up by the red, glaring light; till, towards morning, the fire had driven over the summit of the ridge, and only a great spreading line of rolling smoke showed where the flames were opening a way for them beyond.

At daybreak they started, – the country still shimmered over with smoke, – and the tramp of the sheep throwing up clouds of black dust from the thick bed of ashes below.[53]

By the mid-1860s the days of the heroic droves – stocking the first runs and pioneering the first sock routes – had ended. What these men had collectively helped to create was an industry that by 1867 had generated 27.2 million pounds of wool and injected £1.581 million ($165,635,278 equivalent in today's money) into the economy.[54] But there were still adventures to be had while on the road with stock, and the life of the drover was still rugged, particularly for those who drove long distances. It was an outdoor life that was always going to involve battling the elements, the terrain and increasingly, as motor vehicles became more common, traffic. Skills were gleaned, knowledge and expertise developed, and the role of the drover became more clearly defined.

Who these drovers were, what their job entailed, who employed them and where they travelled will all be covered in following chapters.

Cattle belonging to the Eggeling and Harris families traverse Lake Wanaka on the way to Omakau from South Westland. Date unknown.
Betty Eggeling Collection

2. ON THE MOVE

The drover's role in the emerging farming economy

The outstanding feature of
New Zealand's economic
development is the dominating
importance of the pastoral
industries, which strengthened
as these industries became
more diversified and worked out
characteristic methods suitable
to their new habitat.[1]

The impact of pastoralism and the worth of the 'golden fleece'[2] to the economy of a young country cannot be underestimated, but as a stagnating wool market and overstocked runs took hold in the 1870s, pastoralists faced an uncomfortable reality. Excess stock had little value. Of course, it was the advent of the frozen meat trade that was to be their saviour.

However, refrigeration was not the only factor to ensure the success of this new trade. Also intrinsic to its success was the move to the intensification and stratification of farming. This was enabled, towards the end of the 1800s, by the convergence of agricultural advancement, New Zealand's natural physical advantages, and the Government's political will to subdivide the large estates, together with the purchase of large tracts of Maori land in the North Island. New patterns of stock movement subsequently emerged as the farming of fat and store stock took hold. For the drover this meant employment and the development of relationships with those buying and selling the stock, particularly, the fat stock buyer, the stock and station agent and the trader–dealer. It meant that routes to freezing works, saleyards and farms became well-worn. Ebbing and flowing according to depressions, world wars and the vagaries of the export market, a perpetual wave of stock, often benefiting from days feeding on the 'long acre', was on the move.

The frozen meat trade and the changing farming landscape

By the 1870s the runs were fully stocked with around 9 million sheep, and pastoralists now faced the problem of what to do with surplus stock. Early attempts at canning meat proved largely unsuccessful, and boiling down sheep for tallow was generally unprofitable, especially with the mainly lean merino stock. It was a period where 'the sinister shadow of the boiling-down works, symbol of waste and the crude attempt to save something from the wreck, cast itself over the land'.[3] Some run owners even resorted to driving mobs of culled sheep over cliffs, as they had no worth.[4]

Then came the promise of a solution, with the successful voyage of the refrigerated ship *Dunedin* from Port Chalmers to England in 1882 with a load of frozen mutton. With a limited consumer market for fresh meat in New Zealand and a burgeoning demand in Britain, the birth of the frozen meat trade was timely, renewing the anticipation of wealth to be generated from pastoral pursuits.

Higher production rates were necessary to harness the full potential of this new technology. This meant radical changes to New Zealand farming practices. Changes had already begun before 1882 with the cross-breeding of sheep to create a dual-purpose animal that would provide good-quality wool and meat. Land was beginning to be farmed more intensively, too, as farmers realised the increased carrying capacity of the land when they sowed English grasses, practised crop rotation and applied fertiliser. This process of intensification allowed smaller-scale farms to become commercially viable, a few hundred sheep instead of several thousand sheep were capable of producing a surplus.[5]

Further change came after the fledgling frozen meat industry gained traction. The move away from extensive pastoral farming meant greater demand for land for the 'small man'.[6] Under the Liberal Government's land policy of the 1890s the large pastoral estates were broken up, and millions of acres of land was purchased from North Island Maori, which allowed 7000 farmers to settle on the land.[7] Pasturing tussock country was becoming routine on the plains, as farms in the North Island were chiselled out of bush.

As farming intensified, new patterns of stock production centred on the growth of the fat and store stock markets. This process involved farming effectively according to the country's topography. This stratified approach meant hill-country farming became synonymous with breeding and the supply of store stock, both sheep and cattle, to the lower-lying fattening farms. Young ewes and older cast-for-age ewes from the tougher hill-country farms were sold at large ewe fairs in January and February to stock the lower hill-country farms, where they were bred with a terminal sire to produce lambs that were destined solely for the fat lamb trade.[8] Farms on the fertile rolling hills or downlands and plains focused on finishing store stock, breeding lambs for the fat lamb trade, or stud farming.

A similar pattern emerged for beef cattle. Breeding cows, culled for age, would come off the tougher hill country and be bred for a period on the easier land. Cattle played an important role on the hill country, where their less particular palate improved the quality of grass for sheep.[9] This flow of store and fat stock created cycles of stock movement and a ready supply of work for drovers. Cattle numbers were always fewer than sheep, however,

Roads

The infrastructural development of the small colony was imperative in order for it to prosper. A farming economy needed ready access to markets: roads, bridges and rail needed to be built, and quickly, in order to get exports to ports. Initially, land development focused on areas closest to sea access; however, as demand dictated, further settlement took place in the hinterland.

Building the network of roads necessary for a developing country was a massive task. Roads were at first focused around settlements, then gradually spread out into surrounding farmland. As motor transport became popular, better roading was a higher priority for the Government and improvements were made – although there were noticeable regional imbalances. Taranaki by 1935 had two-thirds of its main roads tarsealed,[10] whereas the thinly populated Far North became known as the 'roadless north'.[11]

Ease of travel along better roads was of obvious benefit to the drover: they no longer needed to negotiate their way through thick vegetation or lose their way in a sea of tussock. As roading networks progressed and patterns of stock movements emerged, stock routes were better defined by local councils. These gave the drover a recognised route to travel, along with the occasional council-owned holding paddock where stock could rest overnight.

In remote areas, of course, there were still basic tracks, such as the South Westland track over the Maori Saddle, that were followed for years to come. Beaches, too, were still used as roads for many years; for example, people living in Paturau in Golden Bay continued to negotiate Westhaven Inlet with their stock for decades.

These sheep heading to market between Stratford and Ongarue in 1910 faced a muddy road – typical of early conditions until roading began to improve. *Sir George Grey Special Collections, Auckland Libraries, AWNS-19100804-10-3*

Railway

The development of New Zealand's railway was imperative, too. The emphasis was on establishing the Main Trunk line in both islands, but where possible branch lines extended into newly developing agricultural land and facilitated the speedy transportation of produce to markets and of stock to other farms, to saleyards and, as the freezing industry developed, to the works.[12] Between 1879 and 1884, railways had already transported 686,287 head of livestock.[13]

The significance of the railway was not lost on the news reporter of the *Timaru Herald*, 7 February 1878.

> MAMMOTH TRAIN: At half-past 5 p.m. yesterday a special train, consisting of 93 trucks, containing in all 5,300 sheep, arrived in Timaru en route from Waitaki to Amberley. It was more than a third of a mile in length, and was drawn by two large tender engines. It left at 6.30 for the north. En passant; what a difference there is between being able to convey 5,000 sheep in one day from Waitaki to Amberley by rail in February 1878, and taking three weeks or a month at least to drive a similar flock in February 1867.

By the close of 1900, despite a deep depression that began in the late 1870s and stretched into the 1890s, some 2205 miles (3548 kilometres) of track was laid; 824 miles in the North Island and 1381 miles in the South Island.[14] Rail networks expanded until 1952, despite progress being stymied by the pre-1900 depression and world wars.[15]

Sheep ready to be loaded onto rail wagons at Turakina Station in 1951. Rail was a huge boon for the farmer as it provided a much more efficient means of transporting large numbers of stock to market. For the drover it meant hours of work loading stock.
Archives New Zealand, AAVK W3493 B1727

A train loaded with sheep passes through Cross Creek Station on the Wairarapa side of the Rimutaka Range, circa 1910. *Alexander Turnbull Library, APG-0148-1/2-G*

Of course, for the drover it meant there was work droving stock to the railhead and loading them into the wagons. Don Monk recalls how he trained his dog to assist in loading sheep:

> . . . this big white dog I had with the black eye – when you were trucking you were pushing them up the race but they would jam in the door all the time, well I taught him to go up and grab the front one by the flank, well it would shoot in the truck and then he'd come back and grab the next one and they'd shoot in and I could load 'em in two seconds.[16]

Although, in all the commotion, one important fact should never be ignored: the door on the other side of the wagon should always be shut. 'Only happened once. By gees you made sure of that.'[17]

Not all animals were receptive to being loaded into the wagons, however. A story is told of some drovers who were having difficulty loading a Jersey bull. One of the drovers, Snowy:

> . . . hit it and down it went dead. And it was a prize sort of a Jersey. And they said, 'What are we going to do now?' and Frank said, 'A couple of you grab it by the horns and somebody by the front leg,' and they towed it into the railway wagon and they propped it up with a bit of hay and sent it on its way. And the bloke said, 'My bull was dead when he arrived, but,' he said, 'he was lying very comfortably.'[18]

as New Zealand beef exports to Britain could not compete with the much larger and closer North and South American markets.

The dairying industry was also taking shape and flourishing, aided by the advent of refrigeration: it had already become a £1 million earner by the turn of the century.[19] For the drover, this meant droving cull cows to sales and boner cows to the works, the dreaded bull droves, and moving dried-off cows to winter pastures. In general, though, the dairying industry never generated as much work for drovers as the sheep and beef industry did.

Ray Stevens, sharing his memories of moving dairy cows from the 1940s through until the beginning of the new century, points out that by the time he started droving, fencing was better, roads had improved, and dairy cattle were a lot more domesticated and easier to control than the cattle that he had seen mustered by what he called the 'real drover'.

> The dairy cattle, she didn't want to go off the road and get tangled, so most of them stayed on the road. And from the early '50s onwards somebody would like to send a mob of cattle away and that increased over the years as people started to increase the size of the herds they were milking. There was a lot of so-called cheap grazing around – an extra cow or two would cover the cost of grazing those cattle. So I would take the calves away, which by then were nearly 12 months old, and for a backload I'd bring back last year's calves which were in calf heifers to the farm.[20]

Ray was also involved with the annual 1st of June Gypsy Day ritual, droving sharemilkers' herds to their new homes. Ray did the work at a nominal rate to give the sharemilker a helping hand: 'I think to shift 200 cows it was something like $8 or $9 a head [by truck] – that was a couple of thousand – and me doing it for $50 a day, it certainly helped them start again.'[21]

The export of pastoral commodities was, of course, a huge boon to the New Zealand economy; for the drover it was their source of employment. The continual growth in trade, particularly in the first half of the twentieth century – which saw Britain commandeering primary produce for two world wars – kept those moving stock on the hoof busy. The depressions of the early 1920s and 1930s slowed growth, but there was always stock on the move.

There were peaks and troughs in the yearly seasonal cycles, with spikes in movement during the freezing-works season[22] and for ewe fairs, cattle weaner sales and the like. Saleyards had their regular sales of store stock, and prime stock for the local market was always in supply; some sales were weekly or fortnightly, others were monthly. Weather and feed availability would result in fluctuations in sale numbers, too. As an indication of just how large the droving industry was, for the period from 1904 to 1913 some 27 million sheep and 30 million lambs were killed for local consumption and export, and 2 million cattle were slaughtered.[23]

The freezing works and the fat stock buyer

The New Zealand Refrigerating Company, established in 1881, was responsible for the first export of frozen meat. It was also behind the building of the first freezing works at Burnside, Dunedin, in 1882. By 1896, there were 17 freezing works up and running throughout the country; seven in the North Island and 10 in the South Island. By 1920, there were 42 freezing works up and running; 26 in the North Island and 16 in the South Island.[24]

As well as the big plants like Burnside and Belfast (in Christchurch), smaller local works opened, such as the Tokomaru Bay Freezing Works. Opened in 1911 by Tokomaru Freezing Co. Ltd, with an initial killing capacity of 2500 sheep and 60 cattle a day, it provided an alternative to a 100-mile drove to the Taruheru or Kaiti works in Gisborne and gave farmers ready access to the overseas market.[25]

Hundreds of thousands of sheep were driven to freezing works around the country annually. These sheep are being driven to the killing floor at Pareora freezing works in the early 1900s. *South Canterbury Museum, 2010/157.05*

Opened in 1882, the Burnside freezing works was the first of its kind in New Zealand. The development of the frozen meat trade was hugely significant for the New Zealand economy; for the drover it provided a steady source of work. *Alexander Turnbull Library, 1/1-019454-G*

As the fat stock industry developed, buyers employed by the freezing works would be out purchasing stock, and there was steady employment for drovers, particularly through the peak of the works season in January. Drovers would move stock to railheads to be loaded and transported to the works; or, where there was no railway, they would drove them all the way on the hoof.

It was a common sight to see mobs of sheep queuing every 400–500 metres down East Street, Feilding, during the works season before stock trucks began to take over.[26] Shepherds in the yards at Borthwicks works in Feilding were sometimes even directed to assist drovers in negotiating the Feilding streets on the way to the works. Horses were kept at the yards for such occasions.

> . . . the boss sent us out to help a drover in. George Davies would bring down 150 cattle or more and we'd go out to the edge of town and help him through town . . . You could bring them down all right, you just had to watch they didn't go into gateways, but there were no real dramas. They were usually fairly quiet by the time they got this far, and most times you just kept on the ball and kept the lead going and kept going in the right direction . . . The drover would just tell me if he wanted me to go in the front, I'd go in the front or the back or

whatever, or watch the sides. We just worked together: everyone knew what they were supposed to do – get the cattle there safely.[27]

Saleyards and the stock and station firms

As towns grew, men with an entrepreneurial bent turned to auctioning surplus stock, sheep, horses, cattle and pigs. Saleyards popped up in towns all around the country, in Rangiriri, Ohinewai and Taupiri, Te Kauwhata, Wairoa, Inglewood, Stony River, Rahotu, Waitara, Okato, Oakura, Apiti, Rangiwahia, Kimbolton, Fairlie, Geraldine, Arahura, Feilding, Addington, Lorneville . . .

As stock and station firms developed, the sales became more regular and formalised. The firms acted as an intermediary between farmer and buyer, and became an important source of employment for the drover. Agents would source the stock from their clients for sale, then organise for a drover to lift the stock from the farm and drove it to the saleyards or to the point of purchase. Some drovers developed long-lasting working relationships with just one firm; others worked freely among various firms, national or local.

In the days of the party line, three short rings was the call for the Sundgren family in Feilding. In the evening, if the phone rang it was inevitably for Charlie Sundgren: it would be the stock agent ringing through with the following day's instructions. Two of Charlie's girls, Bev and Kath, remember feeling they were a part of their father's world when they listened in to the evening call.

> The phone was in the living room so we always heard the arrangements being made for where he had to go, the number of stock he had to pick up and the distance, and he would calculate how many days it was going to take him.[28]

The agents and drovers had a close working relationship. Agents would often make life a little easier for the drover while out on the road by taking out dog tucker to them, or by transporting them to and from the holding paddocks when they were close to home. Some agents recalled how they never much appreciated having to load the drovers' flea-ridden dogs into their vehicles, even if it was in the boot – especially when their cars were new.[29]

Sale day at the saleyards around the country was a place for drovers to congregate and pick up any available work – drafting stock in the yards, shifting stock from holding paddocks, or moving mobs after the sale. The canny drover could earn a day's wages for every stock agent he worked for. Jack 'Boy' Curtis remembered his father working for maybe six or seven stock firms in a day and getting paid by each; 'that's why the old man loved it', Jack recalls wryly.[30] He adds that any drafting work was done on horseback: 'everything was drafted on horse, none of this dancing around in the gumboots in the mud, it was all on horseback'.[31]

Peter Cloake, a stock agent for Dalgety's in the 1950s through until 1970, recalls an incident where drovers at the Feilding saleyards negotiated extra payment for work.

> When they'd finished penning up, that was their break. There was a little drovers' room, in the saleyards, where the officers were (in the old days prior to being burnt down) . . . The old rostrum, in the days I'm talking about, they used to sell the dairy cows through there – the only time it was used was for boners, which were dairy cows that were past their dairying life and they'd come in and be sold as boner cattle so that meat would be boned out and sent over to America for hamburgers and stuff like that. And then dairies and then the rest of the cattle were sold out in the yards . . .
>
> The drovers would retire to their little drovers' room which was basically the room we used to get changed in . . . and the drovers used to sit there and talk, and our head drover was Tom Fitzgerald. Tom was a good bloke – he voted Labour, he was a little bit of a 'Red Fed' – and he used to [say], with certain things, 'This is our time to knock off and we don't start again 'til the sale's finished.' And that was it, and they would have a talk, have a smoke, talk about dogs . . .
>
> When we had the change to rostrum selling that did create a bit of a problem 'cos I think the drovers were paid three pounds 10 shillings . . . it was quite good money and then they would get 10 shillings for every mob brought up from the paddocks and 10 shillings for every mob they took, and I think it was 15 shillings per mob round to the rail and then they would get 12 and sixpence to load a truck, the first truck whether it be for sheep and cattle, and then four shillings for every additional truck . . .
>
> In the '60s, mid-60s about – that's when we had a problem. Then we had to put in additional gates, and then you had to have someone to move them through, of course, and that's when we had to go cap in hand to the drovers and say 'Look we want you to move this cattle: you'll have to get your dogs and you can have your little rest and once 12 o'clock comes we're back into it again.' 'Oh I don't know about that. Oh this will cost you.' And it did I think it cost us another's day pay . . . All of them stuck together.[32]

The annual ewe fairs in January–February were a busy time for farmers, drovers and stock and station agents. Peter Cloake, despite being out of the industry for more than 40 years, rattles off the different fairs that were held: 'There was the two-tooth fair, the old ewe fair, the aged ewe fair, the supplementary two-tooth fair, the supplementary aged ewe fair and the combined ewe fair.'[33] It was not unusual to have queues of mobs waiting to be drafted into pens, and drovers worked hard to avoid box-ups.

Once they were in the yards, not everything went exactly according to plan, as animals all have a mind of their own:

> Another time a dairy cow got out of the . . . little ring at the Manchester end of the yard. There were little gates – if you wanted to get out of the road in a hurry, you could duck out there. Anyway one of these old Jersey cows got out of this little gate and started walking across the road. It was a silly old thing, quiet as anything, and it saw the green grass, I think they were council offices over the road. Anyway they had the main building and then there was a glass passageway into the council chambers, I think . . . that were backed

Manawatu's Feilding saleyards became a central hub for the sale of thousands of head of stock annually. This 1902 image shows hundreds of sheep penned ready for sale. *Feilding Public Library, AGR:me11*

This aerial photograph taken in the mid-1930s shows the scale of operations at the Feilding saleyards. At one time it was considered one of the largest saleyards in the southern hemisphere, with up to 100,000 cattle and 600,000 sheep being sold annually.
Feilding Public Library, AGR:me9

> onto Manchester Street . . . This old cow saw the green grass, the lawn, and so it went over there to sniff around. Then someone says, 'Look over there, there's a cow out there.' So you know everyone starts chasing this poor old cow. Anyway before they could get there she walked straight through the glass, broke it, and then she smelt water, so she made a right-hand turn and there was a small flight of stairs, quite wide, three or four metres, up she went and there was a toilet up there and she wedged herself in – she smelt the water you see – and here she was drinking the water out of this toilet till someone got a halter round her and pulled her out of there.[34]

Over time the sales were centralised and, as a result, many of the small saleyards closed. Allan Barber, a stock and station agent with Wright Stephenson for over 40 years, recalls this happening:

> A lot of those little saleyards went out of existence . . . you see, with motorised transport you don't have to have saleyards here, there and everywhere. In the past, you had to have saleyards within droving distance of stock.[35]

For a young country boy like Laurie McVicar, the saleyards was an exciting place – although he didn't always understand everything that was said.

> One thing I can remember . . . they brought a bull into the ring and it was a milking shorthorn breed that they used to use a lot of – which they've gone away from now . . . we used to get out of school and go down to the sale straight away . . . and they were still selling this day, and they brought this bull in and the auctioneer said, 'Now here gentlemen is a fine milking bull,' and I thought, 'You fool, we all know you can't milk a bull.' I didn't know at the time but it stuck in my memory.[36]

From 1920 to 1923, George McLeod collected stock from local farmers on the Taieri, and drove them in to the Burnside saleyards. In *My Droving Days* he recounts his exploits and describes some of the characters he regularly dealt with.

> On Tuesday of each week I left home at Momona early on my black hack and with Malcolm's three dogs following me I rode to Berwick to Bob Shennan's or old Dumpie as we called him behind his back. He was a good sort for all his rough ways.
>
> Mr Shennan was a small man with a goatie beard, he had two large farms on the Taieri where he fattened cattle. One large holding on the Berwick flats and another two hundred acres at Otokia. He sold his cattle privately to Duke's, the butcher, one of Dunedin's leading butchers for many years.
>
> Mr Shennan was in the habit of selling to Dukes 150 head in one line with delivery of 16 head per week. This was the start of my mob each Tuesday, so after arriving at his home on the top of Berwick hill where I stabled my hack, fed with a full bucket of chaff, my dogs were shut in the stable, I was taken to the house where if it was 9 o'clock, 10 o'clock, or later, the table would be set with a joint of meat, a loaf of bread and ample food I was invited to wire in.
>
> After horse and self had ample food, we, old Bob and self proceeded to the paddocks to draft out the usual 16 bullocks for the road and it was a performance that had to be seen to be believed. Old Dumpie had absolutely no control of his dogs, his bullocks were never handled since the day he bought them from the sale yards or stations. He usually kept his cattle for two years before he sold them with the result they became very heavy and fat and also they were very timid or wild. The following was usually what took place. I was called 'Boy' or 'By' for short.

'By – we'll take four from this paddock', 20 wild bullocks stood staring at us as we entered the paddock – a grunt or two from old Dumpie as he sat on his hack trying to make up his mind which 4 to take, he takes out a plug of black Juno tobacco and chews a piece off the end of it, he never smoked, only chewed the black dirty stuff and after rolling it round his mouth for a time, spat the black ink out which sometimes (if I was on the windward side) went all over my saddle and riding strides much to my disgust.

'We'll take yon two Herefords with the turned up horns – let me see – yon black poll and yon red bullock with the down horns' – so after careful quiet manoeuvring we get these 4 separated from the rest. I slip quietly round, open the gate where one of old Bob's dogs grabs the black bullock by the hock – result, the bullock bounds in the air as if all the devils are after him and quickly bolts back to join the other cattle – come in behind me – 'I'll mall your B-guts' roars old Bob – 'I'll tan your blasted hide'. So, after a 15-minute delay in getting the black bullock back – 'keep that dog in behind you now', I roar to him.

As this photograph taken at the Queen Street saleyards in Upper Hutt shows, anything could happen on sale day – with stock making a run for freedom being a frequent event. The drover needed to be ready for any eventuality. *Upper Hutt City Library, P4-62-787*

'I'll cut his blasted throat, that's what I'll do'. We eventually get these 4 bullocks out on to the road – along a bit further, the next paddock – same performance.

'By – see yon roan bullock, well the one behind him and those two poll herefords – yon blue one and yon white bullock, we'll take' – about 30 bullocks stand staring at us so as we get around them, they all bolt to the bottom of the paddock pursued by one of old Bob's dogs – old Bob digs the spurs into his hack and takes after the dog trying to lash the dog with his stock whip and roaring dreadful oaths – through the cattle and along the hedge gallops old Bob trying to lash the dog with his whip but never quite reaching him. Eventually the dog dives through the hedge and with old Bob roaring at him over the hedge the dog stands looking very crestfallen, so we manoeuvre the correct cattle out on to the road – that's 13 we have safely got, 3 more to get – good we have only been a little over the hour – some days he holds me up for hours – so along to another paddock, I know which three he wants out of this paddock. 'Now, if you drive these 13 along I will slip along and have the other three out on the road ready' – 'right you are' he says – On my own I have no trouble, my dogs know to keep out of the way and only help when I call, that's the lot, at the end of the road he leaves me.[37]

The trader–dealers

The traders or dealers were important clients of the stock agent. These entrepreneurs, who saw money to be made in the quick sale and purchase of stock, were found throughout the country wherever there was stock to be bought and sold. They often worked 'close to the breeze',[38] and were prepared to risk purchasing at one point and reselling at another for profit, seeing the opportunity to make 'a bob here or there'.[39] They played an important role in the stock market – whether they were ensuring continuity of supply of stock for the local market, as the traders on the Taieri did; or providing Manawatu and Waikato farmers – at no risk – with cattle they needed in the spring from Gisborne. And they, too, ensured a steady supply of work for the drovers.

W. L. (Bill) Parkinson, or 'Parkie', as he was known, was a Canterbury dealer buying and selling from Kaikoura down to the Addington saleyards in Christchurch after World War One; he did business in Canterbury until his death in the 1950s. Don Monk worked for Parkie as a lad of about 16. He was called on to do all manner of tasks for him, including regularly droving stock. One job was to meet up with Keith Harrison, who worked at Addington saleyards, along with drovers Jim Watson or Bob French at the Belfast Hotel on a Thursday with the 70–80 head of store cattle Parkie would have bought at the Addington sale on the Wednesday. Don would then drove them on to Parkie's 200 acres at Forrest Ford near Kaiapoi, where they were put into lines ready for resale.

Parkie's annual purchase of Burwood Station's calves, and their subsequent arrival and drove from Kaiapoi railway station, was always the big job of the year. The calves were divided into two mobs, and Don Monk, along with Jim Watson, Bob French and Jim Corfield, would drove them to the cattle yards at the Forrest Ford property. Two days of drafting into lines would follow, along with dehorning. While most of the stock were then onsold, with agents coming in to buy them, 100 to 120 or so were driven six days up to Parkie's property at Lees Valley. The stock would spend about two years up the valley and then be sold or fattened on Parkie's 'Sinai'[40] property. With an eye for the dollar, Parkie would also be out buying up stock around the district. Don would accompany him and Parkie's manager, Les Chaney, on these buying trips, and the next day Don would be off collecting the purchased stock and droving them back to Forrest Ford. He recalls that no money ever exchanged hands between Parkie and buyers and sellers: all the financial transactions were organised through Parkie's stock agent.[41]

Another Canterbury drover at the time spoke of his admiration for Parkie, or 'old Billy' as he called him:

Bill Parkinson's crew in 1953. From left to right: Jim Corfield (farmhand), Bob French (fulltime drover), Les Chaney (manager), Jim Watson (fulltime drover), Don Monk (mostly droving), Barney Brown (rouseabout), Bert Borwick (farmhand). *Don Monk Collection*

Old Billy would head off to Kaikoura or Blenheim – he was a bit of a magician, he'd buy a mob of cattle and one of his boys would take them home, he had three permanents working for him, and he'd see these cattle come through the gate: 'I never bought that cow.' He knew every cow that he bought and he could go back months and say, 'I've never seen that cow before.' He had a memory like a – well I don't know – better than mine![42]

The 'long acre'

Feeding stock along the road verge or the 'long acre' – known as 'the best paddock in New Zealand'[43] – was very much a part of the drover's routine. On a long drove it was the only way to keep condition on the animals. At times, if they were short of feed, farmers might even employ a drover to drove their stock around the district to make use of feed on the side of the road. The judicious use of the long acre could be a real bonus, and there were North Island station owners who perfected its use. The Duncans of Otairi Station, the Hurleys of Papanui and Siberia Stations in the Rangitikei, and Jack Jefferis at Waerenga in the Waikato would buy up store cattle in Gisborne and hire drovers to drove the cattle in the winter months over to their stations. It was to their advantage to allow the stock to winter on the road – it was a cheaper form of transport; and it allowed the stock to feed on the long acre, which gave home paddocks time to 'get away'[44] and be ready for fattening stock in the spring. Alternatively, the station owners could have their farm fully stocked over winter and have stock on the road as well; by the time the road stock made it across to the station, they had enough spring growth to take them on – or they could onsell them and make some money on the sale.[45]

Jack Jefferis's son Sam recalls how they could have several mobs on the road at one time. 'I think the year 83 – 1983/84, I'm not sure which year it was – we had 2700 on the road at one time in five mobs . . . Oh, we were the biggest, the biggest by a long way.'[46] Sam remembers sheep being driven through to their Waikato property as well: the last mob, some 2600 ewes, came through in 1987. He lists the names of some of those who drove for them: 'Keith Ribbon, Captain MacDonald, Jock McRoberts, Bill Pullen . . . Ray Ball, Eddy McVern – they'd be more of the regular type.'[47] As the stock arrived in the Waikato, Sam would spend days on the road moving the stock on to the family properties. There was not much time for him to get away from the Waikato, but Sam seized the opportunity to be a part of one drove across from Gisborne before the droving days petered out.

The Hurleys and the Duncans needed the Gisborne cattle for their

The West Coast goldrush

With the discovery of gold on the West Coast in 1864, a mass migration of humanity occurred. The onset of gold fever in the district brought increased demands on resources. Meat was scarce; and farmers and others in Canterbury saw a ready market for their stock. The Southern Alps were an obvious obstacle to their progress, and so the race was on to find a suitable route. Amuri, Harpers and Browning passes were all used as stock routes, along with Arthur's Pass, of course. Arthur's Pass was particularly busy:

> More than 2800 sheep a month were pattering over Arthur's Pass by 1866. Cattle unsaleable in east Canterbury were fetching £25; wethers worth 18/- to 24/- there could sell for up to 70/-. In 1867 something like 40,000 sheep were driven over to the diggings, mostly via Arthur's Pass. Sheep were also coming in from other New Zealand provinces and Australia, so some had to be driven back to Canterbury in order to revive West Coast prices. For a typical week ending 27 July 1866 the *Lyttelton Times* listed the traffic moving to the goldfields as: males, seventy-one, females, five: horses, thirty-nine; cattle, 198; sheep, 1107. Only thirty-four men and twenty-two horses came the other way.[48]

Progress must have been difficult at times, as Edgar Jones recounts of his drove over Harpers Pass:

> On one trip, going to Hokitika, it came on to snow very hard as we were going over the saddle at the head of the Teramakau River, so that the sheep would not travel, and we had difficulty in forcing them on, and could not get them down to the first small flat, which was the usual camping place. Night coming on we had to pitch our tent in the bush, using the frying pan to scrape the snow off the ground for a place upon which to erect it, and also another place where we lit our fire. It snowed all night, and we had to drive the sheep on again in the snow. It took the whole day to get them down to the first flat, and the latter part of the way we had to tread a track for the sheep, as there was by this time, over two feet of snow. We had again to use our frying pan to clear a place for the tent, etc. It was very cold and snowed all night, and as the hillsides were very steep on each side of us, and covered with bush, trees were crashing down all round us owing to the weight of snow on their branches. It cleared the next day, but there was three feet of snow, and we could do nothing but feed the sheep and horses on the branches of the native shrubs. It was chiefly akeake, a shrub which

> they do not care for. We were two or three days in this place, then when the snow had melted slightly, we trod a track with the horses and ourselves, and gradually drove the sheep down the river. . . . Those sheep, when I got them to the sale yards were no longer fat, and if I remember rightly, I was glad to get three and sixpence per head for them. I could not take them back.[49]

hill-country stations. Arthur McRae was a regular taking cattle down to the Duncans for years. Sometimes he would drove just one mob in a year, other times more. His final drove for them was as late as 1998 or 1999.[50] The Hurleys used the cattle to help break in the hill country by eating the bracken and fern and improving the grass for sheep production. Their land was not suitable for fattening their cattle, however, so they held an annual muster and drove of three- and four-year-olds, which were sold at what was known as the 'Hurley sale' at the Feilding saleyards on the third Thursday in September. This came to an end sometime in 1978 or 1979, when the family bought finishing land and no longer needed to sell their cattle as stores. Stock agent Peter Cloake recalls that the Hurley stock had not been handled much, so they were 'quite lively cattle'.[51] They were driven into the Feilding saleyards in two lines, one via Halcombe and one via Sanson.[52]

Cattle on the road at Matangi in the Waikato district in 1934. The slow unhurried nature of droving allowed stock to graze the 'long acre' as they made their way to new pastures. *Sir George Grey Special Collections, Auckland Libraries, AWNS-19341031-45-2*

Government Lands and Survey blocks

At a time when droving was fast becoming outmoded in the South Island, a regional droving phenomenon was becoming established in the North Island. After World War Two, stock sourced from the Poverty Bay and East Coast region was travelling through onto newly developed Lands and Survey blocks in the Central Plateau region around Rotorua, the King Country and also to North Auckland (later Northland).[53] This was part of a Government push for land development, driven by the need to settle returned servicemen, provide for a growing population and combat a worldwide food shortage. The Government purchased neglected private land and developed vast areas of Crown land[54] into economic farms.[55] Thousands of ex-servicemen were settled on the land as a result of this government policy.

Later, civilian farmers were given the opportunity to take up ballot farms, and hundreds of thousands of sheep and tens of thousands of cattle were needed to break in these blocks.[56] As Barrie Gordon, a well known livestock

An advertisement for rehabilitation farms from *New Zealand Farmer* magazine, 6 March 1947. *Hocken Collections, Uare Taoka o Hakena, University of Otago, S13-377b*

A mob of 2600 sheep in the Waioeka Gorge between Gisborne and Opotiki. Ian Mullooly remembers the gorge as a 'narrow one-lane track' where it was possible to end up with 'buses backing around corners and all sorts of things'.[57] *Archives New Zealand, AAQT 6539 A1478*

manager in Gisborne, recalls: 'I would have up to 10,000 ewes on the road at one time – 3000 arriving in Rotorua, 3000 in the Waioeka Gorge and 3000 being assembled in Gisborne.'[58] From after World War Two through until the mid-1970s, drovers were kept busy stocking these blocks. There was also much stock movement between the various Lands and Survey blocks within a region, depending on the season and the carrying capacity of the land.[59] With sheep moved over the summer months and cattle over the autumn and winter, it meant there was plenty of work for the drover.

The role of the drover may have started out slowly in the early years of New Zealand's establishment, but as the country's agricultural might grew, the networks of freezing works and saleyards developed and farming patterns emerged, the drovers became increasingly important to the pastoral industry. The trader–dealers, fat-stock buyers and stock agents all provided employment, and with ever-increasing stock numbers there was plenty of work for those keen to be on the road in the coming decades.

A flock of sheep being driven along the road in Otira Gorge. Date unknown. *Sir George Grey Special Collections, Auckland Libraries, 35-R1008*

3. THE LONG AND THE SHORT OF IT

Droving routes of New Zealand

Old Jack Flower had done quite a bit of droving. He was in the pub one day and he was having a few ales and reminiscing a bit, perhaps a little bit of a skite, and he said, 'I think I've driven sheep from North Cape to the Bluff.' And some wag at the back said, 'How did you get on across the strait, Mr Flower?' And he said, 'I didn't, I went round.'[1]

As the humorist opposite suggests, there are few areas in New Zealand that would not have had stock moving on the hoof through their environs at some stage. From the Far North to the Deep South, from the high-country runs or the plains of Canterbury, the rich west-coast farmlands or the rugged hill country of the East Coast and even from the offshore islands, stock was on the move.

Determined pioneers carved out a living for themselves from bush or plain. Any land was seen as a potential farm, no matter how isolated, and successive generations continued the work of their forebears to ensure the rapid development of this small island nation. The Government built roads, bridges and railways to open up vast areas for farming. There was work for drovers in all corners of New Zealand.

Stock was moved on the hoof in all corners of the country. This picturesque scene from 1928 shows sheep being driven through Kaihu district, some 120 miles (193 kilometres) northwest of Auckland city. *Sir George Grey Special Collections, Auckland Libraries, AWNS-19280405-39-3*

Regional droving patterns emerged throughout the country. In the high sheep-producing areas of the east coast of both islands and as far south as Southland there was plenty of droving work, particularly where there was no rail system. In areas such as the West Coast and the Nelson region, where stock numbers were lower, there was much less work and consequently fewer drovers. Often the drovers working throughout these areas were trader–drovers or farmer–drovers.

In an isolated community like Karamea there was no regular drove. Long-time resident Karl Jones recalls that it could be six months to a year between droves. Drovers would come in and collect up prime cattle and cull cows and drive them out to Westport. Karl smiles at the memory: 'It was always fun to sit on the fence and watch this droving go past.'[2] It was an event in a small community: 'There were no fences anywhere, there was bush everywhere. Stock would regularly break from the mob and be thinking about home. Then there would be a real to-do with dogs and fellas on foot. If they didn't succeed they'd maybe get them next time.'[3]

In the North Island, Taranaki became synonymous with dairying: it produced sheep and cattle, but not in the same numbers as in other regions, so there was less work there for drovers. Elsewhere, particularly in the hill country such as the Rangitikei, as bushland was cleared and developed, sheep and cattle production increased and there was plenty of droving work. Manawatu, with its central location, became a pastoral hub; and the Feilding saleyards was once one of the biggest in the southern hemisphere.[4]

The 'local' and 'overseas' drovers of Gisborne

'Local' and 'overseas' drovers is how the Gisborne and East Coast drovers were known from the 1950s through until almost the end of the century.[5] Gisborne stock and station agent Barrie Gordon organised more than his fair share of these 'local' and 'overseas' drovers, particularly those heading into the Bay of Plenty. The 'locals' drove mobs into the Matawhero saleyards, worked in the yards during the day, then moved mobs back out after the sale. Gisborne and the East Coast was good breeding country for both cattle and sheep but had little fattening land, which meant there was plenty of store stock for sale. Sheep and cattle were bought in large quantities and driven to where they were needed around the island. This generated work for the 'overseas' drovers.

The 'overseas' drovers were divided into the 'northern' and 'southern' drovers.[6] The northerners took stock into the Waikato, Bay of Plenty, Central Plateau and even up into Auckland.[7] The southerners moved mobs down into the Hawke's Bay, Manawatu and Rangitikei regions. While the drovers may have considered changing their routes, in reality most stuck with their well-trodden paths. As 'southern' drover Arthur McRae, a drover for some 50 years, points out: 'I never went north . . . you knew your job, and you knew you had contacts . . . you had to have paddocks, somewhere you could leave a lame beast and that sort of thing.'[8]

Transporting stock around the waterways

An aspect of any droving day might include loading or unloading stock from trains, boats or ferries. In the early days the coastal fleets, or 'mosquito fleets' as they were known – generally smaller vessels, often under 100 ton – were a vital means of transportation and communication before communities were linked by road and rail. Farmers relied on these vessels to transport their livestock. The fleets worked around harbours such as Auckland and Lyttelton, calling in at remote bays; others plied the coastal

Isolation was no deterrent to getting stock to market. These sheep, bound for the freezing works, are being loaded onto a small steamer on the Whanganui River in 1902.
Sir George Grey Special Collections, Auckland Libraries, AWNS-19020710-11-5

The paddle steamer *Antrim* plied the waters of Lake Wakatipu from 1869 until 1905, transporting sheep and cattle to and from stations in the area. The PS *Mountaineer*, TSS *Ben Lomond* and, of course, the TSS *Earnslaw* (known as the Lady of the Lake) all worked the lake at various stages. Station owners relied on the steamers to transport their stock to market. *Hocken Collections, Uare Taoka o Hakena, University of Otago, S12-222c*

The *Antrim* discharges sheep beside Lake Wakatipu. *Hocken Collections, Uare Taoka o Hakena, University of Otago, S12-222b*

waters, crossing river bars and travelling inland to river ports.⁹ Some worked Cook Strait, transferring livestock between the two islands. As road and rail developed, however, the coastal fleets declined.

Transporting stock in this way did not always go well, as an excerpt from W. H. Heays's diary reveals. Heays was working on the Auckland Steam Packet Company's wooden paddle steamer the *Coomerang,* which was transporting cattle from Gisborne and Napier.

> Sometimes we took cattle to the Thames, beached the steamer and, at low water, discharged the cattle on the beach, night or day. Some of the cattle were very wild and in shipping them, they broke through the race and onto the steamer's deck. Then there was a scatter of everyone that was on the deck, anywhere out of the road of the beast, myself included. On one occasion, I cleared along the deck through the alleyway and on to the low poop deck with the beast after me.

> On the poop deck was a long glass-top skylight. I ran round that
> to get out of the beast's road and the beast tried to cut me off. He
> attempted to jump over the skylight and landed on the skylight with
> his hindquarters hanging down into the Saloon and stuck there. A
> headrope was got on him, a tackle got aloft and he was hoisted out,
> and put down the hold.[10]

Allan Barber, stock agent with Wright Stephenson Co. from 1942 to 1985, recalls picking stock up from around Lake Wakatipu and loading them on the *Earnslaw* before there was road access. 'It was all a matter of gates, they used to have a heap of these gates that you loaded on and made them into pens, and then the stock was all loaded into the pens. They used to do cattle as well . . . but then they just hosed the boat down and put the passengers on and away you'd go.'[11]

Island farms

Droving stock out of isolated spots throughout the country always brought unique problems, but there were even greater challenges for those who farmed the offshore islands. Great Barrier Island, D'Urville Island, the Chatham Islands and even the subantarctic Campbell Island were all farmed. The Campbell Island venture failed and the sheep there eventually went wild, but the other islands all successfully shipped stock to market on a regular basis – and still do.

Stock first had to be mustered, then driven to a wharf or bay and loaded onto a boat for the mainland. D'Urville Island had flat-bottom scows that could come in on the tide and be loaded, then go out on the tide. Today, cattle and sheep are still barged between D'Urville Island and Havelock in the Marlborough Sounds. Stock are run along a temporary causeway across the beach and onto the barge, where they are penned. When they reach Havelock they are run off the barge and into stockyards at the marina. From there they are loaded onto stock trucks.[12]

Totara Flat farmer Pat Kennedy remembers as a boy going up to D'Urville Island, where his father purchased sheep and had them driven down to their farm.

> They punted the sheep to Havelock and they were driven from
> Havelock down to here. They would come through the Wairau Valley
> and through the Buller Gorge and down to Inangahua Junction and
> out through there.[13]

Up until less than a decade ago, stock on Great Barrier Island was driven onto the wharf and loaded on board a ferry that transported them to Auckland, where they were driven onto the wharf and then loaded onto trucks. Today they are loaded onto trucks on the island and then onto the ferry.

In the case of the Chatham Island stock, before a suitable wharf was built, sheep were taken out to a waiting ship on surfboats and hoisted on board in slings. Cattle suffered a similar indignity – they were made to swim alongside the surfboat. With two boats working full-time, a dozen cattle could be loaded in an hour. Considering the cattle were mainly wild, standing up to 6 feet high with horns and killing out at 900–1200 pounds, the job had its hazards!

Sheep are unloaded off a boat in the Marlborough Sounds, seemingly none the worse for wear from their sea voyage. Date unknown. *Alexander Turnbull Library, PAColl-6304-05*

Until a wharf was built on the Chatham Islands, sheep bore the indignity of being transported by surfboat to a waiting ship and hoisted on board in slings. Cattle were towed behind the surfboat and then hoisted in slings into the hold of the waiting ship.
From Chatham Exiles: Yesterday and To-day at the Chatham Islands *by Frank A. Simpson, Reed, Wellington, 1950*

A drove might stretch out for days or weeks; or it might be short and intense, requiring great concentration and strong nerves. Owen Gibbens recalls driving stock out of the railhead in Greymouth in the 1940s.

> In the town of Greymouth, for instance, the stock route was originally centred on the wharf because of the shipping, then on the rail as it was up until my last association with it. The rail siding in Greymouth is at the western end of the town, and we have the Grey River on the north side of it, and the end, if you like, on the west side. You find that the rail siding is wedged right at that point, so you have the river on your left and a lagoon behind you, you have a series of railway lines between the river and the saleyards, the tracks could be six or eight wide, then we had a set of good sheds and then a wharf behind where the dredge was berthed in the lagoon area. It was a tight area to get cattle that had hardly seen a man before under control.
>
> You had the challenge of controlling your horses in this area – it wasn't a good place to be tarsealed . . . So out they would come, the gate would open and out would go the stock and you'd always get the one that's away and the ride is on. This was a place where man realised how good, how essential the great animal the dog was . . . So you've got the drive underway and you're holding the stock and a decision might have been made earlier as to whether we went down through Greeson Street, the start of the stock route, or will we take them across the reclaimed ground. And if they were really bad it would be the reclaimed ground and then that would depend whether the tidal movement was in or out or just holding. If the tidal movement was on the rise it was a sight to see – you see the roaring of rivers in mountainous areas – well the water would come through there with great force, and we maybe, depending on the stock, have to take the cattle through that way so that we could get them under control sufficiently to carry out the rest of the drive.[14]

Iconic droves

Three iconic droves in New Zealand's droving history were the South Westland cattle drove, the annual spring Far North cattle drove, and the Richards brothers' drove from way down the Turimawiwi River in the Tasman district.

The South Westland cattle drove from Okuru to Whataroa on the West Coast was carried out by local farmers and 'ring-ins'; walking their stock out from their isolated farms was the only way of getting them to market. Ken Lewis offered farmers in the Far North a cheaper alternative to trucking store stock to fresh pastures. And the Richards brothers Edwin and John (Ned and Jack) carved a niche for themselves in the backblocks of Tasman district: for years they drove fattened cattle through to Appleby for Alex Thompson, the butcher. Nothing about such droves was easy – no matter where in New Zealand they drove, the terrain and weather presented challenges, along with the inevitable battle of man versus beast.

The South Westland drove

The national tourism poster of the South Westland cattle drove captures not only the magnificence of the environment, but the importance of the drove to our droving history. In a day and age where it is possible to drive comfortably from Hokitika to Okuru on the West Coast in a matter of hours, it is difficult to comprehend a time when nothing but a track connected the two settlements; and the track, though it passed through some of the most dramatic scenery in New Zealand, presented considerable danger and challenge to travellers.

The corner of the world where this drove commenced was an unlikely spot for cattle raising, but a tenacious assemblage of families had carved a niche for themselves out of what can only be described as the 'wild west'. These survivors of the ill-fated Jacksons Bay Special Settlement, established during the Julius Vogel administration in the 1870s, took to raising stock, particularly cattle, along the river flats from the Haast River down to the Cascade and Arawhata rivers. They made a living from the annual or biannual stock sale, spring and autumn, depending on cattle numbers. Life was tough – irregular shipping of supplies and inadequate roading meant that the families were cut off from the rest of the country – and they became resourceful and enterprising.

There was no wharf, and shipping stock out of the area was difficult, so they began making regular droves northward – initially a one-month trek to the Arahura saleyards just north of Hokitika, some 200 miles (322 kilometres) away.[15] The route was by no means new. Settlements from Paringa north, established during and after the goldrushes of the 1860s, meant that a road of sorts existed, much of it along the beach. The Haast–Paringa stretch, a particularly rugged bushclad track 4–5 feet wide, followed what was once a Maori trail that had been developed as a packtrack for diggers, avoiding difficult coastline around Knights Point.[16] The drove was shortened by some 70 miles (112 kilometres) in 1913 when cattlemen in the area started a sale in Whataroa. In 1915 saleyards were officially opened.

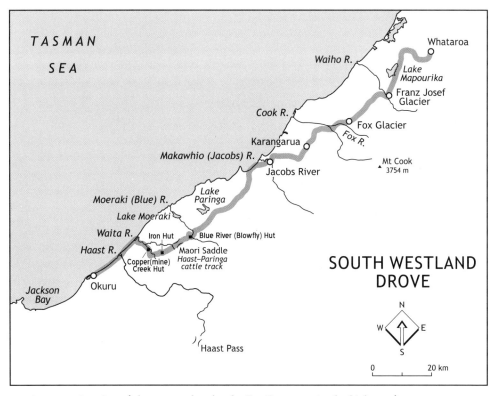

An approximation of the route taken by the Far Downers. As the highway from Hokitika made its way down the coast, the drove grew shorter. In later years the drovers travelled from Okuru to Paringa, and cattle were then trucked to Whataroa. The last drove over the cattle track was in 1961.

Mustering stock on river flats and bringing them down to home paddocks was the first order of business. Each family in the community would be mustering their own stock with ring-ins to help where necessary. Nolans' herds were the biggest in the district, and they often had extras in to help. Tony Condon recalls experiences as a young man mustering on the Nolans' Cascade and Arawhata runs in the 1950s. The horses carted up to the Arawhata were then ridden in to the Cascade. It was the speed of the operation that left a lasting impression on Tony. He says of Des and Bill Nolan, 'when they travelled it was like boy racers and it was all go and when you got there your horse was buggered'. It was always Tony's intent to never be bucked off one of the 'fresh' horses, to spare himself a ribbing from the older men.[17]

Run cattle were not much used to human contact. Although they might have calmed down by the time they got to Whataroa, Tony Condon recalls that while mustering, the cattle were so wild 'they'd run right through them'.[18] Good horsemanship was essential, and there was no room for error.

Cattle are loaded onto the MV *Gael*, which regularly travelled along the West Coast delivering supplies to small coastal settlements. In Betty Eggeling's recollections, her husband Charlie's attempts to transport cattle by boat were largely unsuccessful.[19]
Betty Eggeling Collection

Des Nolan's riding skills again impressed the young Condon: 'Des Nolan could ride a horse like no one.'[20]

Before they crossed the rivers on the droving routes, they had to face their own rivers – and like any South Westland rivers, there was always the threat of flood. Allan Cron, helping on a Nolan muster, recalls the fright he got crossing a flooded Arawhata.

> Another time, it was in the Arawhata, we brought out 225 bullocks from Cascade and [it was] pissing down with rain, absolutely poured – eight hours from Cascade to Arawhata Bridge used to be, and we decided, the Arawhata River was high but we'd swim the cattle because they probably wouldn't face the bridge. We made the fatal mistake to swim them, and we did actually get them over as it ended up, but I was on a little mare, Twinkle. And I was on the downstream side of these cattle, trying to force them up – once again stepped over the edge of the fall and she wasn't a big horse, she was only a small

horse, and she rolled over and I got my feet free of the stirrups. I remember coming up and seeing the grey water in front of me – into face and raincoat and leggings and all this shit on, and saw the stirrup flying through the air and I just drove my hand into it, and God it took all the skin off my bloody hand . . . and I just hold on to the stirrup and she towed me out. That was a lucky bloody escape.[21]

Betty Eggeling (née Buchanan) moved to Okuru as a young girl in the early 1930s. She loved nothing better than to be out working stock, and was known to be 'a bit of a horsewoman'. For Betty, working with her husband, Charlie, mustering stock up their Waiatoto run was always a joy. They would work right up to the foot of Mt Aspiring, about 30 miles (48 kilometres) up the Waiatoto River. It would take two days to get up there. The hut was halfway up, so they'd drive the stock from the top of the run down to the hut and stay there and then muster from that point. One trip Betty recalls still being out at 11 o'clock at night. All she could do was follow the sparks from the horse's shoes on the rocks: 'the sparks that came off his horse's shoes on the rocks was telling me whereabouts he was . . . you just couldn't see. His horse was dark, and my horse was dark and the bush was dark.'[22]

Any recalcitrant stock could always be treated to a little bit of homegrown ingenuity, as Allan Cron recounts.

> Years ago we were driving cattle down to Haast, this beautiful November spring night we got to the five mile, now there's been 100 acres cut away, and it was this scrub country and was all these niggerheads and bungies and all that and the smell, the smell of the plants and the scent just beautiful in the late spring evening after the sun had gone down. Anyway this bloody bull parked up, he only had to go another half mile and he was in the holding paddock – bastard parked up – what were we going to do? We got the bloody dogs in there; no, they couldn't shift him. He had his arse backed in under a big log, there was no way he was going to shift. The old man said, 'Hold my horse,' he says, 'I'll shift the brute,' and he gets down and pulls off a great big armload of dry ferns and he gets under the log and he pokes it under the log in between his back legs and then the ole wax match. It flared up, of course, and then it just roared into life 'cos the old bull he gave a couple of bloody kicks and Christ he's off. And all that you could smell right down to the holding paddock was singed hair. But that shifted him.[23]

Whatever the challenges, stock would reach the home paddocks for a final opportunity to gain condition before the long drove. The families in the meantime prepared for the trip, shoeing and resting horses, preparing food and swags.

Over the years a pattern of travel emerged for the drove: travel distances, accommodation and holding paddocks were all considerations. Families with smaller mobs banded together to help each other. Nolans with their bigger herd often went alone. Buchanans and Eggelings teamed up often with the McPhersons, Cowans or Harrises. Crons might go alone or join the Nolans. Two or three mobs might be heading away with staggered starts so the mobs would not cross. Those on the drove would stay in huts on the route, hotels, or with family and friends along the way. But although accommodation may have varied along the way, there was no variation in the route. The drove, starting at Okuru, went across the Haast River via the beach and then on to Copper Creek, over the Maori Saddle to the Moeraki River (or 'the Blue', as older locals still refer to it), along past Lake Paringa to rest at Condons' property, on to Jacobs River then to Karangarua, then further on to Fox Glacier and northward to the Waiho River, around the shores of Lake Mapourika and finally arriving at the Whataroa saleyards. The ease with which they can rattle off this route belies the degree of hardship that those on the drove experienced.

It was no mean feat to cross the West Coast rivers, especially as they were so prone to flooding. The Waiho and Fox rivers were glacier-fed – Betty recalls seeing lumps of glacier ice in them at times. Along with the rivers there was a climb to around 1800–2000 feet (550–610 metres) above sea level on a track where stock had to walk single file, and where drovers had to pay special attention so the stock did not get between the bank and their horses. There was considerable risk of animals pushing them off the track, and the fall could be from a considerable height. Adding to that were the stroppy run cattle, some four- or five-year-olds with horns, and the distraction, when on the main road, of service cars passing; tourists taking photos; and of course the West Coast cold, torrential rain or even a bit of snow.[24] It is little wonder that those who went on such droves won the respect of the Coasters and took on almost a legendary stature.

Tony Condon remembers the admiration he felt as a boy when he watched the mobs go past.

> I can remember from probably before I went to school these mobs of cattle . . . They were bushmen, sort of, musterers. They were, I guess – Banjo Paterson's verse from 'The Man from Snowy River', this was what it was about really. I mean these guys were from Snowy River and the horsemanship of some of these fellas was unreal, absolutely unreal.[25]

Remembering the Droving Days

The old time drovers of yester year have long since gone to their rest,

Through a life time of toil when times were hard,
they always gave of their best.

With sturdy hearts and a will to win, they never gave up the fight,

For surviving was the name of the game,
and they strove with all their might.

Their deeds have stood the test of time,
and showed us all the way,

With camping out and mountain tracks,
they started the droving day.

The next generation of stockmen grew up,
to learn the skills,

To break young horses and ride them,
learn the hard way by taking the spills.

For a stockman's horse was his treasure,
and on him he had to depend,

To carry him over the countless miles,
and looked upon him as friend.

Of mustering and crossing high rivers,
and there were so many ways,

To be tough and ready for action and to carry on the droving days.

But times had changed by the sixties,
when cattle trucks came to the fore,

When the long weeks of droving the cattle trails ceased,
and were no more.

A drover's dogs were his faithful mates,
and followed him day after day,

Hard working and honest they had a hard life,
but never failed to obey.

When memories of droving the Paringa track gently recede in a haze,

The last of the drovers of days gone by,
remember the droving days.

—Bill Nolan, November 1997[26]

Each stage of the drove, whether it was the miles to cover or the rivers to cross, had its own challenges. However, without a doubt it was the Paringa or Maori Saddle stretch that elicited most comment from those who drove the route. Des Nolan describes it in *The Far Downers* as 'A sidling road and slips and narrow tracks and creeks. You know, it was a bit of a hill-billy thing in a way.'[27]

Myra Fulton (née Buchanan) remembers:

> That was a big day, and if it was wet it was so miserable . . . You'd get in your cut of stock. The first cut takes 12, and they push them along to keep the rest coming along. The next cut was normally my cut and I'd have about 20. And you'd just keep them moving all day. But you couldn't stop, you couldn't do anything, you couldn't get off your horse or anything because they'd panic. They'd never seen anybody on foot. They'd seen people on horses but not on foot . . . You couldn't talk to anybody, but you could hear the dogs away behind with the people bringing up the rear, and the one in front, well you sometimes could call out to them but not very often . . . Didn't worry about anything to eat . . .[28]

For smokers the day could seem extra long, as rolling a cigarette with wet, cold hands was a bit of a battle. Myra well remembers the kindness of Bernie Cowan, who left a rollie in a twig for her to grab as she passed.

Patience was needed for stock as they went at their own pace, and in a stretch like this you could not put pressure on them. Mickey Clark recalls going up over the Maori Saddle.

> . . . it was the longest day, well as I said you had to stay on the horse, the cattle were a bit flighty . . . In some ways it was quite frustrating sometimes you'd be on a narrow bit of road and you'd be going down into these dips into the creeks and one beast would think, 'Well I'll stand and have a look around and have a drink.' You're too far away, you couldn't do anything about it and you'd just have to sit and wait. And when it moved on the others would move on and when they went you could go. You couldn't force it, it could be 30 head of cattle away sometimes, and it could be the lead one decide to have a drink and a look around, but you just sat there quietly.[29]

Two features of greatest concern along this 17-mile (27-kilometre) track – 'Chasm Creek' and 'Slippery Face' – were spoken of with a certain degree of respect and, for some, perhaps even fear. Slippery Face was a stretch of track that, as its name suggests, was on the move. It was the bane of the roadman's life, and any major rain could set it on its way. Those crossing it on horseback were always aware of its instability and of the drop below.

Chasm Creek, near the top of the Maori Saddle, presented different challenges:

> [At] the head of the Whakapohai and the creek where it crossed the track was called Chasm, and [where] you went across it was a waterfall that came down – I mean a waterfall, a young river depending on how much rain there'd been, and it went down, it boiled and then down there was probably about 150 feet and I saw and heard a bullock go over. It stirred me . . . you heard it bellow and then thud . . . The waterfall was probably not that high, 50 or 60 feet or something, it boils down, the more rain the more water. A terribly bad place for dogs . . . we used to catch the dogs and put a rope on them and tow them across so they didn't get washed across.[30]

In fact Bill Nolan remembers seeing his uncle Paddy Nolan's dogs swept over the edge when pushing cattle through the 'raging torrent' of a flooded Chasm. That trip a mob behind them got stuck two days at the Iron Hut waiting for the water levels to drop. It was the only time the Whataroa sale day was delayed to allow the Crons', Harrises' and McPhersons' stock to get through.[31]

It was not all doom and gloom. There were moments of light-heartedness and humorous events along the way that offered relief from the monotony:

> It always caused chaos if we met tourists on the track. They were inclined to dispute their right to the track, but I thought it was quite entertaining. In fact, sometimes it was a real laugh. The cattle had never seen anyone on foot before. They were reared up the valleys and some were still clean skins. One tourist wanted to stand on the track and photograph the cattle as they streamed by. I do declare that anyone could have heard Dick bellowing almost from Copper Creek, 20 miles away . . . Some of these tourists thought they were hidden in the bush, but the mob smelt them and spooked, charging past in twos and threes at a full gallop, not giving time for any photography.[32]

Some with a mischievous streak might set things going despite being warned of the consequences. Betty Eggeling was one such character:

> It was along the Maori Saddle that the telephone line, our one telephone line ran . . . The telephone line often went from tree to tree, not from telephone post to telephone post. It was driving along there that often

you could touch it as you went by . . . and anyway I was driving along there one day, we each had our cut of cattle. This day when we stopped – it must have been at the Iron Hut – we stopped for the night then when we started off the next morning, I said to Charlie, 'I think I might hit that telephone line with the handle of my whip to see what the reaction would be.' Because with all the cattle strung out along . . . when you get a mob of cattle like that one little noise and it goes all along.

Charlie told her not to. 'So what do I do, I hit the wire with the whip handle, set the whole mob off.' Betty laughed at the memory of Charlie's reaction, adding, 'I made off that I didn't hear.'[33]

Whatever the challenges they faced, the aim was to get the stock to the sale looking as good as they could. Sale day in small communities was always a big day. The stock were drafted up and penned before the sale in turn by all those who had travelled – some locals, some from north of Whataroa, but mainly those from further south – creating a great hubbub. There was a sense of anticipation as to the sale prices: it would be a welcome injection of cash for people who led a very basic existence. 'It was often a rainy day for the sale,' local Eric Kennedy comments. 'They used to say the rain would put another dollar on a beast. A beast looks better when it's wet.'[34] The inward flux of buyers, sellers, traders, stock agents – and drovers, of course – meant a busy few days for the local publican, as well as an opportunity for news and gossip to be swapped.

Betty Eggeling was not involved in drafting stock, but she would watch the sale; and she made the most of the opportunity for some female company – not something she got a lot of at her isolated spot in Okuru.[35]

Whataroa residents still recall the sale days of their childhood. School was forgotten as they looked on – but never from the rails of the yards; it was too dangerous with the horned cattle.[36] The refreshment hut – a small wooden building with a counter and a copper inside to boil the water for the tea – served cups of tea, pies and sandwiches. It was always well attended, and it was an opportunity for local women's groups to raise some funds. Mary Kennedy, a local, laughingly recalls, 'We sold lamb sandwiches before lunch then after lunch we called them hogget.'[37] The stock agents supported the women's groups by giving items such as a box of tea to raffle off.

The hotel was the centre of post-sale proceedings, along with the auctioneers' car boot shout. Both were largely male domains; the evening was filled with yarn telling, singing, sale-price confab and prank playing. Tom Condon (Tony Condon's father) once rode a big cream mare into the bar; and there were stories of undoing horses from their gigs and running the shaft through the fence then hooking the horses back up again, or taking the horse

Richards Bros cattle at Patons Rock Beach, circa 1960. Ned Richards is on the white horse. The drovers have misjudged the tide, so stock, drovers and dogs are left to get wet. *Pat Timings*

Bill Pullen with a mob of cattle at Whakamaru in the mid-1970s. They are heading to the Waikato. *Bill and Gill Pullen Collection*

Farming families on Great Barrier Island have no alternative but to barge their livestock off the island. Up until 2000, the Mabey family's stock were transported in pens on board the barge and loaded onto trucks at the wharf in Auckland. *Mabey Family Collection*

Stock trucks now travel across on the barge to Great Barrier Island and stock are loaded at the farm – a faster, more efficient method of transporting them. *Mabey Family Collection*

Bill Pullen with his young son Roy on Hawai Beach, Bay of Plenty, in the mid-1970s. Droving took on quite a different nature when you were on the road with your family – as suggested by the nappies on the line. *Bill and Gill Pullen Collection*

The Pullens stop for a smoko break on Traffords Hill in the Waioeka Gorge, Gisborne, and Gill seizes the opportunity to dry some washing. The mob is on the way to Jack Jefferis's farm in Te Kauwhata. *Bill and Gill Pullen Collection*

Longtime North Island drover Arthur McRae on the road with a mob of cattle. *Arthur McRae Collection*

out of the shafts and putting him back in facing the wrong way.[38] There was a real hooley if prices were good; less so if prices weren't so good. Mary recalls in later years that if prices weren't good people would head home pretty quickly.[39]

For those from the deep south – the 'Far Downers' – the trip home was much quicker. With just themselves, their horses and dogs to worry about, the distances travelled in one day were far greater and they could be home within three days. With backsides 'like the pad of a dog's foot'[40] they travelled home to take up the work waiting for them.

The Ken Lewis Far North drove

In Broadwood, January 2008, a gathering of old drovers from the Far North came to recount memories of what was an institution in the region – the Ken Lewis spring cattle drove. Ken, a cattleman born and bred, took to droving as a lad when he was presented with his first pony by his father. Despite having one wooden leg as the result of a car accident at a young age, Ken developed a reputation for being a cattleman of note and owned several farms in and around the Kaitaia area. As a lad he took to mustering the likes of Te Paki Station up in the Far North, and droving boner stock monthly to Moerewa freezing works. This gradually developed into regular droves, taking cattle from the Far North down to the Kauri saleyards near Whangarei and sometimes as far south as Wellsford. But it was the annual spring drove that he took over running around 1945[41] that became a Far North institution and captured the hearts of everyone who participated.

Into the 1990s the cattle still marched from Paua to Whangarei. The over 200-kilometre trek took five to six weeks. Drovers worked long hours in basic living conditions. The drove was seen by local men and women as an opportunity to break with routine, and they left their regular pursuits to participate in it. Ken offered a contract service to buyers, many of whom were from the Waikato and King Country: he charged a flat fee per head to Kaitaia or Whangarei. The cost of accommodation, holding paddocks, drovers' wages, and the price of lost stock were all covered within the fee. It was a cheaper alternative to trucking, and offered the buyers several weeks of free grazing on the long acre, giving their home paddocks opportunity to ready themselves for the new arrivals.

The starting point of the drove had altered over the years from Te Paki Station to Thoms Landing, as land ownership changed; but by the mid-1970s Paua was the starting point for the spring sales. From the Paua sale, the mob built from sales of mainly Lands and Survey stock at Te Kao, Cape View and Onepu stations and Houhora, the smaller mobs merging with the larger mob as they travelled slowly down towards Kaitaia. Drovers were charged with the responsibility of staying with the main mob while Ken went off to sales to contract with new buyers, and often to pick up some cattle for himself.

From the 1950s till the 1970s Ken could be moving as many as 4000–5000 cattle in two or three mobs, staggering the droves so as to limit inconvenience to motorists. One mob would follow State Highway 1 through the Mangamuka Gorge all the way down to Whangarei, avoiding Kaitaia. Another mob would travel right down to the end of Ninety Mile Beach and over the Herekino Gorge into Broadwood, pick up more cattle at the Broadwood sale and then carry on onto the highway at Mangamuka. A third mob picked up at the Peria sale on the east coast would be driven through Otangaroa and join up with one of the other mobs at Mangamuka, or travel on its own to Whangarei.[42]

In later years there were fewer cattle on a drove, so they travelled as one mob. From the first sale in Paua until the last in Houhora a week passed, and Ken's mob in the finish was paddocked at a Government block at Sweetwater, ready for the long march to Whangarei where they would be drafted and trucked south.[43] From there, the route went through Herekino and into Broadwood, onto the highway at the Mangamuka–Broadwood junction, down through Rangiahua, up the hill to Okaihau and then, avoiding

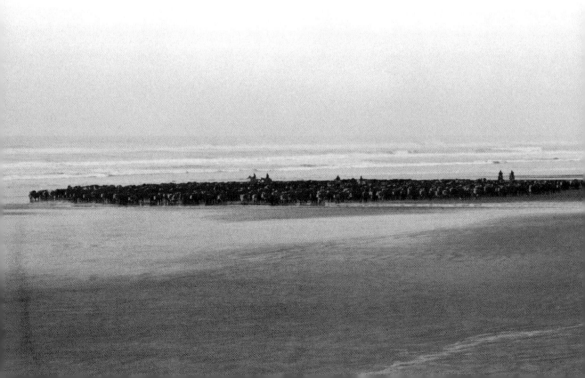

Ken Lewis's stock are driven out along Ninety Mile Beach in the Far North in 1981. While the scene looks peaceful, even on a good day it was cold and the sand could blow harshly in the drovers' faces. Northern Advocate/*Mike Hunter*

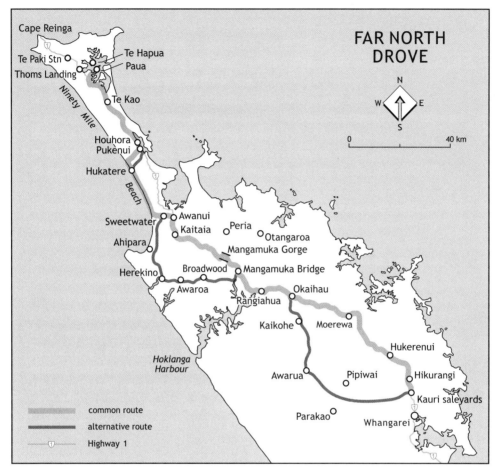

An approximation of the route taken on the Ken Lewis drove, which was an institution in the Far North. The stock usually went down State Highway 1 through to Whangarei. In later years, it was common for drovers to take the alternative (western) route to avoid the tutu in Mangamuka Gorge.

main roads, on to Kaikohe through Parakao and the Pipiwai–Mangakahia route through to the Kauri saleyards near Whangarei. This route avoided the Mangamuka Gorge, where tutu grew – a native plant that is lethal to stock.[44] The route entailed hours of toil, battles with unpredictable cattle – and the just as unpredictable traffic – and, of course, the weather.

Planning was an essential element for a successful drove. Phyllis Clements recalls sitting down with Ken prior to the drove, planning meals, holding paddocks and accommodation.[45] Over time Ken replaced packhorses with a truck, which was in constant use throughout the drove delivering food, carting tired dogs, transporting drovers to accommodation at the end of the day's drove, and transporting his horse. Stock was drafted

A drover watches over stock as they graze near the Hukatere lookout in the Far North in 1981. Northern Advocate/*Mike Hunter*

off along the route as buyers communicated their needs, and by the end of the drove the numbers of drovers would have dwindled from as many as ten to maybe three. The logistics were huge, and considering the value of the animals travelling on the hoof there was no room for error.

Ken didn't have it all his own way in the 50-odd years he drove. For a short time in the early 1980s Kerry Coulter, another local, set up in opposition to Ken, offering buyers an alternate drove of smaller mobs down to Whangarei.[46] Allan Crawford, a Rangiahua farmer, needed to supplement his income in the difficult farming climate of the mid-1980s, and also took to droving the route for several years. Allan drove his cattle further south to Paparoa, which gave stock owners a longer period to ready pasture. The drove became unsustainable for Allan as stock sale patterns altered in the region and numbers of cattle at the sale decreased.

For years it was predicted that the Ken Lewis spring drove must surely come to an end – newspaper articles wrote of it possibly being the last one as early as 1979 and into the 1980s[47] – but still Ken took it on, despite increased traffic. He did not do it for the money; according to him it was no goldmine. As he put it, 'Really I do it more because I like the work. It's not always easy work either. It's long hours. It's just that I *like* handling cattle.'[48]

The demise of the drove was taken out of Ken's hands by another accident in 1991, which left him in a wheelchair – and so the days of the Far North spring drove came to a sad and abrupt end.

Mo Moses and Bobby Wells, both older Maori men, go back a long way with Ken. Both recall taking stock down the Te Paki Stream onto Ninety Mile Beach and droving them right down the beach to Hukatere in the days when the sale was based there. Bobby talks of the days when you could take your horse out into the surf, cast a hemp line and catch a snapper for lunch. They both remember month-long musters with Ken on Te Paki and Te Hapua stations before the sales. Bobby speaks with admiration of Ken's enthusiasm for handling the wild, tough run stock.

The stories unfold of the terrain and the environment – the spring weather, the mud, the push up the Mangamuka Gorge to avoid the tutu – but without a doubt the most picturesque stretch of the drove was the early section travelling down Ninety Mile Beach, starting at Pukenui and coming out at Sweetwater. Ken himself would have known the route well; he first drove cattle down the beach in 1933.[49] The panoramic *Country Calendar*-style images of hundreds of cattle stretched out across the beautiful northern beach, the sand dunes and the stockmen on horses whistling their dogs immediately come to mind. The drovers were not above seeing the beauty: on the one hand, they waxed lyrical, with a hint of emotion:

> One of the things I remember I was bringing a big mob out over the sand hills over the back of Cape View and the sun just coming up and – very, very pretty to see them coming out over the sand hills.[50]

But then they would add a note of realism:

> but a southerly wind blowing and sand in your mouth every time you whistled your dog – and in your eyes. Get your sandwich out at two o'clock in the afternoon and try and smuggle a bit in your mouth and it's got sand on it . . . even on a nice day it's cold, and that sand blowing over your face is really something else.[51]

Not everything always went according to plan, but the drovers learned to take the good with the bad. If they were unable to extricate themselves from trouble, there was always the ability to deny all knowledge of any untoward event. Neville Clotworthy found it worked a treat.

> We'd just got down off the sandhills, I think there was five of us with this mob of cattle . . . three of them disappeared pig-hunting – they

saw some pig tracks, they were gone. And the fella out in the lead –
he was quite young and I yelled at him and I told him he didn't need
to go pig-hunting. And I thought, 'Oh yeah we're getting on pretty
good,' – then I saw a bullock with a pine tree hanging out the side of
his mouth either side and another one over there with a big mouthful
of pine trees – and a big flat that had just been planted with pine
trees the day before – they unplanted them a helluva lot quicker than
they'd been planted.

Next day Cassidy [Ken was known as Cassidy] came along to me
and said, he said, 'Did you run into some pine trees up there?' I said,
'Yeah, yeah we came out through a few pine trees.' He said, 'Did you do
them any damage?' I said, 'No, no I never saw any damage.' So he went
back to Forestry and said, 'No, no these fellas said no they didn't come
out through any pine trees – it must have been somebody else.' And
Ken said, 'Did you come down through pine trees?' a couple of days
later. And I said, 'Yeah, but they were only little pine trees, Ken.'[52]

If denial was not to work, the drovers knew that the buck ultimately
stopped with Ken, and the drovers were not averse to leaving the explaining
to him. John Holland found this tactic useful when facing the wrath of a
disgruntled farmer.

The next episode was Victor Yates and Albury Hobson. We were on
the road with a big mob of cattle – 600 head. We got to Rangiahua,
and Victor and Albury – chasing girls. Got up next morning – no
Victor and no Albury. And old Ken says to me, 'I don't know where
those boys are. Don't know where those boys are, but we've got to
get on the road. You take these cattle, John.' So he let the 600 out
on the road and away we go. We just get along . . . by Gillies. And
Gillies, this is way back and Gillies is a big strapping fella about six
foot two and about 16 stone. And I've got this horse – about 32 years
old – he's going to shoot it for dog tucker on the road, and I'm riding
this thing – it can do about one and a half ks or miles an hour. And
I'm taking these cattle, nothing you could do, they're just 600 head
of cattle going – no one in front and I was just taking them along . . .
And old Gillies had a single-wire electric fence right around his house
and a paddock off to the right. And, of course, these cattle were really
hungry and they just walked over that no trouble. . . . and I had a real
good dog, a Smithfield cross dog and he was a real top dog and I just
put him around and he brought these cattle round in this big wave
and they just flattened the fence – it was history. And I got around and

> here's Gillies standing out the front and there's steam coming out his ears! And here I am on this broken-down old horse and he's into me.[53]

John relates how, with a bit of quick thinking, he placated Gillies and avoided a beating: 'Hey, hey, it's really not my fault that bloody Victor and Albury are supposed to be here helping me but they're off chasing birds and I'm left with the cattle. Ken's the boss, you see him.'[54]

It was a rugged life, working long hours – days often started at three or four in the morning. The hours were cause for comment from Bill McCready, who joined the drove as a lad; his father Selwyn had driven for Ken for years. After finishing a day's droving with a few quiet beers at the Mangamuka pub they had headed back to Ken's house at Kaitaia around midnight. After tea, and shooting and hanging a horse for dog tucker, all headed for bed. Bill goes on:

> Got to bed and next thing there's banging around, 'Get up boys.' It's three o'clock and we're sitting at the breakfast table having our breakfast and this fella seated at the table playing with his sausage and eggs and Ken says, 'What's wrong with the food, boy? Tucker not good enough?' 'Too bloody close to teatime.' It had only been two hours since we'd had tea.[55]

During the gathering there is a sense of camaraderie as stories flow over a glass of beer. Admiration is voiced for Ken's strength and ability to get a job done. There is also a sense of having shared in a unique piece of New Zealand's history – and there is an almost tangible sense of nostalgia for days long past.

Ned and Jack Richards's drove

Brothers Edwin (Ned) and John (Jack) Richards took on the development of land at Paturau, in the northwestern corner of the Tasman region, in an era when carving a living out of isolated and rugged blocks of land was a common feature of New Zealand farming culture. The land, originally purchased by their father in 1898, was divided between family members in 1910, and the brothers took to clearing bush and establishing pasture. Fiercely determined and hardworking, they took on a further 550 acres of bush at the Turimawiwi and continued to buy and clear land when they were able.[56]

Breeding cattle and sheep was all part of the brothers' activities, although Ned's son Harry comments that they were mainly traders,[57] and it was as traders that they became known around the Nelson district. Operating

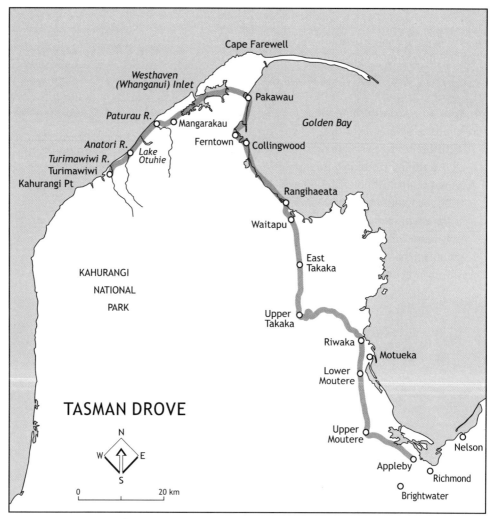

An approximation of the route taken by Ned and Jack Richards along the beaches and roads of the Tasman district. Over the years, the route from the Turimawiwi River to Appleby became well worn.

under the name of Richards Bros, the men made money while they developed their land by buying up store cattle from around the Collingwood and Nelson area, as far south as Tapawera and as far afield as D'Urville Island. The D'Urville Island stock was driven to a family block at Ferntown; the rest were driven to the Turimawiwi for fattening. Under an arrangement with Thompson Butchery that began sometime in the 1920s, the brothers would deliver around 90 head of cattle every six weeks to Alex Thompson's grazing land at Appleby for the next 34 years.[58]

The well worn route went from Turimawiwi to Paturau, along Westhaven (Whanganui) Inlet and the Pakawau beach to Ferntown, through to Collingwood and along the beach to Rangihaeata, turning inland to Waitapu. At Waitapu, Ned would leave Jack to carry on with the rest of the drove while he headed home to begin work readying the next mob. Jack would get assistance from locals to help him along the rest of the route, from Waitapu via East Takaka through to Upper Takaka, over the hill to Riwaka and Upper Moutere, then the final leg through to Appleby.[59]

The drove took 12–14 days all up, including rest days along the way at Ferntown and Waitapu. The actual droving time was seven days:[60] the first day from Turimawiwi through to Paturau was a half day; the next day from Paturau through to Ferntown was a big day – timing the tide well to take the cattle along Westhaven Inlet and along the Pakawau beach, they would start often in the moonlight to take advantage of cooler conditions and less traffic on the road.[61] The biggest push, however, was the stretch up over the Takaka Hill to Riwaka. If the cattle were big it would take two days.

Jack had many ring-ins who helped over the years. Barry Grooby recalls being roped in to help by his father Lance, a World War Two veteran who drove around the Motueka area up until the 1990s. Barry remembers as a lad of only 10 or 11 going up Takaka Hill in the 1950s. The job was quite daunting for a young boy out in the lead on a large horse followed by horned cattle – some with a 1-metre horn span.

> My only thought was: how am I going to get to the top of the Takaka Hill and get that gate open so these things with these great antlers can get in there and not attack me? We got halfway up the Takaka Hill to the horseshoe hairpin bend and got a fair way round the hairpin bend and I hear a voice down below me yelling and screaming at me that I should never ever have been that far in front of these great bullocks. So I had to stick with them and all I was left with was this sitting on top of this great big horse with this stockwhip – cars and that coming and you had to work out very quickly whether you let the stock go so the cars could get through or whether to hold them up. For a young fella of probably 10 or 11 it was ... yeah ... At the time it was high-stress stuff but I look back now and just laugh.[62]

Kath Morgan (née Cook) remembers being called on to help Jack, as she was a keen rider and there were not many riding in the area at the time. Heading away up the hill from her Riwaka home at 5 a.m. she would meet up with Jack and the stock as they came down the hill. She was quick to say that while there was not much traffic on the shingle road at the time,

The enduring partnership of Tasman district identities Jack and Ned Richards saw them working well into their later years. Top: Jack sits astride his horse Stormy. *Kathleen Morgan Collection* Bottom: Ned Richards out riding with his dogs. *Bill Richards Collection*

'some of the bends and things were awful'.[63] If the stock were going to Lower Moutere the drove would be six or seven hours. She would usually be out in front of the mob making sure gates were shut. Some of the cattle, Kath recalls, were six-year-olds with horns that had come from the far side of the Kahurangi lighthouse. Nobody would want to tangle with them, and the 'poor people who met them on Motueka bridge on their bikes had to get right up on the kerb to get out of the way . . .'[64]

Kath remembered Jack as 'a good man'; but at the same time, she recalled, 'he was a very big, strong man and he didn't take anything from anybody . . . he did what he wanted to do . . .'[65] If a car owner barged into the mob, Jack would certainly give them 'their pedigree'.[66] After the cattle were delivered to Appleby, Jack would do his rounds to buy up more store cattle to take back to the Turimawiwi, so the cycle was repeated.

Harry Richards, Ned's son, is still farming and buying up land around the area with his son today. He has memories of going on the drove when he was young: he and his father would be out the front of the mob while his Uncle Jack 'dominated' from the back.[67] He recalls that the two brothers would shout and carry on at one another: '. . . you know, they'd say "get your bloody dogs out of there" and all that sort of thing . . .' However, despite this, they had 'an enduring partnership'.[68]

The family is unsure of exactly when the brothers stopped droving, but their nephew Bill Richards comments that his uncles 'just about died on their horses'.[69] The two brothers died just a year apart, Ned in 1968 and Jack in 1969. Whatever their foibles, Ned and Jack were undoubtedly tenacious pioneers in the Tasman district, working hard to make a living and creating a legacy for future generations. The cattle droving that they began died with them.

Over 20 years have passed since the Northland drove came to an end, and some 50 years have elapsed since the South Westland and Richards Bros' droves were a part of New Zealand's droving tradition. By the time the Northland drove came to an end, however, the twilight days of droving were fast approaching, and to see men and women out on the road with stock was becoming a rare sight.

Over many decades, all over New Zealand, stock had moved on the hoof – from short distances to epic droves to the saleyards or freezing works. For many people, the work was their livelihood; for others it was just another element of country life.

This statue of a drover and his dog was erected in Feilding in 2003 and acknowledges the significant role that drovers played in the local economy.
Keith Dobson

4. THE FACES OF DROVING

The drovers on the road

Unfortunately fellas that shifted stock all went under one heading 'drovers' – but there're drovers and hoolers.[1]

When people reminisce about the droving days a rollcall of drovers is named, often with a remembered quirk or character trait. There's Joe McLaren – or 'retread Joe' as he was called – who was known to make soles for his boots out of car tyres;[2] and Linc Campbell – 'a big strong man, a champion dancer . . . a good fighter too – quick on his feet.'[3] Smokey Thompson and Ken Harding – they both had 'tin whistles as big as thumbnails always in their mouths – never saw them take them out even when eating – beautiful whistles, beautiful tone'.[4] The 'barefoot boys', Dave and Mike Satchwell, who hooked their boots over their saddles because they preferred to walk barefoot.[5] Captain MacDonald, 'an old Maori chap [who] drove around the coast to Gisborne'.[6] There was Jim Watson in Canterbury, 'one of the best drovers I've ever seen – he could make a blimmin cow talk'.[7] One drover was even known to have a bottom lip 'that could hold a fortnight's rain';[8] and another was remembered for his lightning pace: 'I'd never seen a joker that could walk so slow and keep moving like he could'.[9] Jack Esler was remembered at his funeral as 'a rough diamond'.[10] Others included Alby Burney, Billy Riddle, Scottie Moyes, Keith Sowerby, George Davies, Ian Mullooly 'the singing drover', Sam Esler, Bill Pullen, Arthur McRae, Jack and Boy Curtis, Sonny Osborne, Arthur and Elliott Grieve, Reg and Llewellyn Win, Harold Traves, Bill and Max Husband, Len Bell, Alex Cocks, Jack Kane, Albert Weenink . . . Whether droving was their life or just a necessity for a period of time, the individuals on the road added colour to rural life. Ian Mullooly recalls:

> I used to do a lot of singing, a terrible lot of singing . . . They called me the singing drover in the finish. No, it's a fact, I'm not kidding . . . I had a mouth organ with me . . . Yeah, I was always singing, I was a happy sort of fella – opera or comic opera or tragic opera . . . [11]

Gender, race, class or age were no barrier for those moving stock on the hoof. You might say it was an equal opportunity occupation – although

A group of sheep drovers in 1937, relaxing with their dogs during a welcome lunch break. *Sir George Grey Special Collections, Auckland Libraries, AWNS-19370421-49-1*

perhaps many of the women and children 'roped in' to assist on a drove would argue it was more like an opportunistic employer. An assortment of individuals could be seen out 'on the road' with stock at any given time.

From the pastoral era through until to the twilight years of droving in the late twentieth century, drovers came from a variety of backgrounds and experience. Some were well turned-out, their horses groomed and their dogs well fed; others were rough and ready, with use of vernacular to match. Some were teetotal; others inhabited the local as often as they could. There were those who liked the ladies – and some who sired children along their route. There were undoubtedly some cowboys in the mix, opportunists who saw easy money to be made, who neither recognised nor honoured the craft of droving and who sullied the reputation of drovers. For the most part, though, those on the road took great pride in their work, with emphasis on ensuring stock arrived at their destination in the same, if not better, condition than when they left. Often, school had been anathema to them; they were far more interested in life outside the classroom, generally involving horses, dogs and stock. In a sign of the times, many started their working lives as young as 14 or 15, often drawn into it by the example of their fathers.

The job required tenacity, patience, integrity and a good deal of fortitude, and for anyone who demonstrated such traits, work was assured. Sonny Osborne, a Feilding drover in the 1930s, recalls his local working environment:

> In 1930 there were 40 drovers in and around Feilding, this being during the depression years. Some were top drovers employed consistently, but many, less efficient, worked on an average of one

Bim Furniss, droving around Huntly in the 1940s and early 1950s, well remembers his father's offer to quit school and take up droving: 'He didn't have to ask me twice.'[12] Rob Murray, of the Far North, resorted to wagging school in order to get on the Ken Lewis drove: 'I used to go to Herekino School . . . as I got older I'd ride my horse to school, wait for the cows to come and tell the teacher I was going to the toilet and out the back and on the horse and on the road.'[13] After Rob had an altercation with a teacher in the fourth form, however, Dollar Murray, Rob's father, decided that perhaps more learning was to be achieved on the road than in the classroom and removed Rob from school. While Rob was excited at the prospect of droving there was perhaps a slight hesitation when he heard Dollar talking to Ken on the phone, telling him to 'get all your mongrel horses . . . We're going to stick this little prick on them.'[14]

Bill Pullen was not much enamoured of schoolwork as a Correspondence School student in the backblocks of Gisborne; he would rather be fencing. He would go and visit the fencers on the place. 'I used to zoom down there with my Correspondence, and while they were having smoko they'd do my Correspondence and I'd dig a few holes for them. It was a fair trade, I'd get my work done and they got their holes dug and everyone was happy.'[15]

He was not above getting one over on the Correspondence teacher, either. 'I had quite a few tricks up my sleeve. I wrote to my teacher and told her I was terribly ill and she wrote a lovely letter back, 'Dear Billy, Don't worry about your lessons and make sure you get well again my dear.' So I waited for three months and then started doing my lessons again. In the meantime I'd broken in a dog or two and [was] generally having a good time.'[16]

Drovers on the road in 1930. Left: When the right spot was found, drovers seized the opportunity to boil the billy and take a break. Right: Restored after a brief rest, the drover is ready to get back to work. *Sir George Grey Special Collections, Auckland Libraries, AWNS-19300409-48-2*

or two days a week. Allowing for the cost of dogs, feeding same, a paddock for a horse, horse shoeing, maintaining a gig and oilskin clothing, one wondered how they managed to keep a wife and family on a pound a day when work was available. These drovers were handicapped because they had not enough work to train and steady their dogs.[17]

Gentlemen drovers

During the early pioneering period necessity dictated that all manner of men rubbed shoulders. Runholders, generally educated and moneyed, who were to become the social élite in the newly founded colony, worked alongside shepherds, labourers and roughnecks in order to see their stations established. Unshackled from the strictures of Victorian society, the runholders, with the enticement of expanding wealth and the promise of adventure, embraced the rough lifestyle of early runholding, of which droving was very much a part. However, as these gentlemen settled themselves and their families on the station it became common practice to employ a manager to oversee the running of the station, and time was freed up for other occupations, particularly involvement in local or national politics. Droving for them was simply an early necessity and certainly not an intended occupation.

The rural worker

At the heart of the early droving story is the rural worker. Often referred to simply as the 'shepherd', or the 'man' or 'helper' in the early diaries of landowners, the rural worker – often itinerant – took on whatever agricultural and pastoral work was available to him in order to provide for himself and his family, if he was married.

These were hardworking men. Alfred Duncan looked back over his working life in the young colony in *The Wakatipians* (first published in 1888) and noted that there was no thought of protest or strike:

> We used to work far longer hours than the labouring class has to do in Great Britain; we camped out and ate whatever food we could get without complaining; indeed, had the early settlers of our colonies been as fastidious as the lowest working men are now-a-days in Britain, the progress made by these colonies would not now be the admiration of the world . . . But in the pioneering days we had no holidays, neither had we any thought of an eight hours movement,

and Sunday only differed from other days in the week by the
appearance of a 'plum-duff' on the dinner table.[18]

As John E. Martin argues in *The Forgotten Worker,* the generic term 'rural
worker' is more appropriate than giving a more specific title, as the fluidity
of movement between roles makes it difficult to categorise them.

> In the nineteenth century, at a time when both pastoral and arable
> farming was labour-intensive and carried out on a large scale, the paid
> rural workforce was vital . . . Some central waged occupations in the
> arable sector were general farm hands and labourers, ploughmen,
> haymakers, harvesters and threshing-mill hands. In the pastoral
> sector they included station hands, shepherds and boundary-keepers,
> musterers, drovers, shearers and shearing-shed hands. There were,
> however, many other important casual and seasonal jobs related to
> the land such as land clearance and bush felling, sowing of grass-seed
> and crops, grass-seed harvesting, rabbiting, fencing, and so on. People
> moved from one occupation to another according to circumstances
> and the season, so that it is often unrealistic to categorise them by the
> specific job done at a particular time. The broad generic term 'rural
> worker' often describes them better than a particular occupation.
> . . . Shearers might also take on small-scale farming or other work
> on farms or stations. General farm hands might try their luck at
> harvesting or other seasonal work for higher wages. Small farmers and
> farmers' sons might spend the summer supplementing their income by
> harvesting or shearing.[19]

The rural worker had to be versatile; and those who found permanent
employment on a station were expected to be willing to tackle a multitude
of jobs, including droving. Such versatility is represented in the accounts
of William Morrison, a 40-year-old Ayrshire labourer, and of William
Wyatt. Morrison was originally brought out by William Deans in 1839
as a carpenter, and he showed his capabilities as an assistant surveyor,
carpenter, guide and drover. Later he became manager of two stations,
one of which was 'Ready Money' Robinson's Cheviot Hills estate.[20] Wyatt's
signed employee agreement with Mt Peel Station in 1856 shows that he was
expected to be versatile – and to take on droving, if necessary, as a part of
his terms of employment.

> I agree to become your servant for the term of one year from the 21st
> April 1856 and will employ myself in any work I am directed. As I

> understand something of Carpentry I have no objection to rendering every assistance in my power in the way of sawing timber for building etc . . . I have no objection to drive sheep and will travel to any part of New Zealand and render my best assistance in helping them along. In the above service for the year I shall be happy to remain receiving the sum of 50 pounds.[21]

There were also some unlikely candidates among those workers, as Sir John Hall, New Zealand Premier 1879–82, recalls in a Canterbury Jubilee publication:

> Sheep driving in those days, like travelling, was apt to make strange bedfellows: some of the best helpers to be obtained were the rough characters, who, as whalers or runaway sailors, had been long in New Zealand. They were not choice in their language, but they had great local knowledge, took hardships without grumbling, and were honest workers if grog was not too handy. They did not generally go by the names given them at their baptism, but by others more appropriate and descriptive. Amongst my early friends were 'Long Tom Coffin', 'Yankee Sam', and others, especially 'Billy the Bull', who was a genial, honest, hardworking little fellow when sober, but very apt under the influence of liquor to get into trouble.[22]

Children drovers

Age was no barrier to gaining experience droving; in fact, many a drover's story begins with being plonked on their pony as a child and told to lead a mob. Children were expected to shut gates, block side roads, slow traffic and ensure the mob stayed behind them. Children as young as six or seven might be seen ahead of a mob;[23] and girls as well as boys went 'on the road'. Rural children tell stories of unloading the sheep off trains and guiding them home after school, or driving cattle to the weekly sale early in the morning.[24]

Childhood experiences of droves are etched in the memories of many. John Sullivan of Franz Josef recalls at an early age watching the cattlemen of South Westland bringing their stock through for the sale at Whataroa.

> It was a feature of my life, you know, twice a year there used to be big numbers of cattle come from south going up to Whataroa around October and March. I always had to be woken up to watch the cattle going past . . . It was a big thing for me, always had to get up and see the cattle go.[25]

Left: Carol MacKenzie as a young girl on her horse. Carol took any opportunity to be outside working with stock, and delighted in assisting the drovers who passed through the family property. *Carol MacKenzie Collection* Right: A young Julia Sowerby is pictured with her father, Keith, on a drove in the Manawatu. *Julia Sowerby Collection*

Although he never worked on a South Westland drove, John was involved in droving family stock. He remembers his first drove in the late 1930s:

> First time I went droving I was probably about 10 or 11 – Dad jacked up a job for me. The cattle were leaving here to go to Ross. We had a paddock over at Franz and the usual thing was to go over there, leave them there for a day's spell and then go on the next day. On this occasion they must have been going on a Saturday, I suppose, because I got this job to go over there. I thought I was made, to get this job . . . I was put out in front of the mob. Dad told me first day out they're always pretty fresh, so going up the first hill here give them a bit of room . . . My uncle had a cut, about 15 to 20 . . . I wasn't scared or anything, but I was thinking I've got a pretty responsible job here. That was my first introduction.[26]

John's first taste of droving left him wanting more. He did get to go on quite a number of droves before trucking took over in the area; their family last drove stock in 1946. Like many others, he was sorry to see droving come to an end.

Laurie McVicar remembers the loneliness and monotony. As a young lad of only six or seven, he even nodded off in the saddle when out in front of a mob.

> Oh yes, I used to get that job and I used to hate it because I often didn't have a dog and I'd be on the pony and they'd ask me to go in front to take cattle over bridges like at Ahaura or through the township. And I . . . just after we'd got on the road . . . I'd get sent out the front and those cattle would get used to the look of the back of my pony and they would follow him. And it helped put cattle over bridges like Ahaura River and through the township – but it was so damned lonely – there might be a couple of drovers behind, but there I was stuck out there. And a boy on a horse in those days was as good as a man – and possibly sometimes better. But oh God, it was lonely.[27]

He passed the time by making whip crackers.

> Oh well if you saw a flax bush with some healthy looking flax and you had a pocket knife, you might cut a piece of flax and make some whip crackers as you went along . . . it was just something else to do. No one had any fancy telephones or radios or anything like that. And you'd be like that for hours and hours and hours.[28]

Despite the monotony, Laurie always jumped at the chance to drove around the district. He loved riding and working with stock – although he ended up taking up an apprenticeship in engineering, having been warned he 'couldn't make a living chasing cattle'.

Labour shortages during the war led to young Brian Philps being roped in to droving.

> Well it was during the war and I was probably about 14 . . . and the neighbour up the road, I used to help him in the shearing shed, and then they wanted stock taken over to Shannon from Foxton because they had a farm over there. So I borrowed Dad's dog and someone else's horse and away I went. I didn't mind skipping school . . . just drove on my own . . . only small mobs of dairy heifers, things like that . . . I know I enjoyed it and everything seemed to go well – we got them all there.[29]

Droving was not just for country children, either: lads in town were only too willing to seize opportunities to assist drovers by guiding stock through the streets. Malcolm Grayling and his older brother, Phillip, were town boys; their father was a solicitor in New Plymouth. In the 1930s and 1940s, Fitzroy, New Plymouth, was still very much a village: the boys grazed their ponies on vacant land, rode to school each day, and played cowboys and

Indians down at Fitzroy Beach. Their neighbour, local stock dealer Mark Barnett, took the boys under his wing and initiated them into the world of droving. Under his tutelage, Malcolm, who was only 11, was out on his pony with his dog, droving stock with his older brother around the New Plymouth district, negotiating trams and cars as they passed the Fitzroy shops on their way to the Waitara works. They helped Barnett with droving for four or five years, and the experience was influential in their later decision to head into a life of farming.[30]

Some children seized any opportunity to be out with stock; others, though, approached it with some trepidation. When Barbara Kenton was growing up, her father, Keith Sowerby, was always out on a drove around Feilding, where he was a well-known and respected drover. At 15, when she was old enough to drive a car, Barbara was roped in to help out on the odd short drove with her father. One drove in particular sticks in her memory.

> I remember we used to go up, say, to the Cheltenham holding paddocks, bring a mob of cattle down to the corner of Kimbolton Road where the Cheltenham hotel is, and he used to say to me, 'Now you just take this mob of cattle down the road – the two dogs they'll look after them, they'll stop at all the cattle stops so they won't go in . . . someone else's gate . . . You just drive down slowly behind them, I'm just going up the road to get another mob of cattle and I'll bring them down.' Now we'd have these two mobs of cattle: one here, I'm controlling it with the two dogs, and him bringing another one down . . . And I can remember that in this particular mob that was behind, there was a bull in it and it decided to stop and talk to some cows, and I was absolutely terrified of this thing. So, when it decided to come and catch up with the rest of the mob I remember getting into the back seat of the car and pulling a rug over myself for fear it would come up to my window or something . . .[31]

With cattle going down all around her with tutu poisoning, poor Bronwyn Matthews in the Far North, when she was only seven or eight, was asked by her preoccupied father to hold a staggering beast's tail in the hope it would prevent it from going down. Her father, Neil, ruefully recalls:

> I remember she had on a brand-new pair of shoes and I cut the ear on a cow – she was pretty wonky and I said to Bronwyn, 'Here, hold the tail straight up in the air,' I said. 'Might be able to stop it from going down.' Cos when they went down they were dead. Anyway so she stood there hanging onto it for a while, then the next thing the cow

backed up and Bronwyn went flying down into a drain that was full of water and that was about the end of these brand-new shoes. I wasn't the most popular person, believe me.[32]

Maori drovers

Maori were very much part of the droving scene, particularly on the East Coast, the Central Plateau, and in the Far North. The role of Maori in the country's pastoral and agricultural landscape has long been recognised.[33] As the pastoral runs were taken up, the skills of young Maori were noted – in particular in shearing and riding. As historian Herries Beattie comments, 'The younger natives learned to ride and break horses and to act as stockmen. On horseback the young Maori was at home, and soon began to equal, if not excel, his pakeha comrades.'[34] Sir Apirana Ngata also noted:

> . . . the Maori may be classed with the casual workers of New Zealand. We owe to his labour much of the pioneer work all over the North Island, as bushman, roadmaker, packer, fencer, and boatman. In the pastoral areas he has excelled in shearing, as shepherd, stockman, and drover . . .[35]

The development of much Maori land into productive farming units after World War Two, and the large numbers of stock moved onto developing Lands and Survey blocks in the middle of the twentieth century, meant there was plenty of droving work for those who wanted it.

Women drovers

With the legacy of the hardworking, stoic pioneer women behind them, many rural women have incorporated droving into an already demanding life. Women in the pioneering years accepted the role of following and supporting their husbands, whatever their ventures; and with the uptake of pastoral runs, some in the most isolated of regions, it was not uncommon for women to be seen moving 'lock, stock and barrel' with their men. Barbara McCreath, newly married to John Hay of Pigeon Bay in 1858, spent her honeymoon travelling from Pigeon Bay to their new home at Tekapo Station. The 170-mile (273-kilometre) journey took a month with their six-bullock-team dray, which they slept under at night, three horses and 20 head of cattle.[36]

Women figure in Census data as drovers after World War One through until the 1950s; and there were many more who assisted when needed –

often out in the lead regulating the speed the stock moved and ensuring they did not wander down side roads or into other properties.

Where they had the opportunity, many rural women jumped at the chance to get out on the road with stock: often it was seen as 'an outing'[37] and a break from daily routine. Christina Morunga of Broadwood recalls droving with her husband, Ben, from Mitimiti around to Broadwood in Northland right up until 2008: '. . . you meet people along the road and have a yarn. Some people have still got time to stop – tourists certainly do. And it's just nice going along. And there's a real satisfaction in getting the job done and getting the cattle there in good condition or better sometimes than you'd picked them up . . .'[38]

Mary Stevens would drive in her little Mini out in the lead of a mob for husband Ray, when moving mobs of dairy cows in the 1980s:

> It was wonderful, it really was wonderful. I would have thought it would have been the most boring, boring job – you know, sitting in the car from six o'clock in the morning till five at night or whatever time – but it wasn't. There was so much to see, and you got to know the cows, the lead cows. And there were lots of little adventures that made it wonderful.[39]

Although she is quick to point out one unpleasant element:

> One thing I did discover was if you're way the heck out in the country and you really need to go to the toilet there is just nowhere in New Zealand where there isn't somebody way up on the hill or down in the valley – you know. Dangerous thing, by crikey.[40]

While Mary was out the front of the mob, she would hang out Ray's oilskin on a signpost or something if it looked like rain; 'or at times there would be a piece of paper tied and fluttering and there'd be a note, "buy 50 acres here".'[41]

World War Two brought a shift in roles for New Zealand women throughout the country. With the mass migration of men to the war there were severe labour shortages, and women were 'manpowered' into all manner of industries, including farming. Women, no matter where they hailed from, city or country, were stationed on hundreds of farms throughout the country. They worked hard to ensure agricultural production was not disrupted – especially with increased demand by Britain for agricultural commodities – and that regular farming routines were maintained. Many of these 'land girls' had to drove stock to saleyards,

Land girl June Matthews holds a weak lamb while on a drove in Mangaorapa, Hawke's Bay, in 1943. The labour shortages brought about by World War Two saw women taking on all manner of roles in the workforce – and droving was one of those roles. *Alexander Turnbull Library, 1/4-000726-F*

railheads and freezing works. In Dianne Bardsley's *The Land Girls*, the women recount their exploits; and many of them recall droving experiences.

> With dogs she had trained herself, a land girl was often expected to drove up to 1000 Merinos or Corriedales along metal-surfaced roads. 'I had to sleep in the Rakaia riverbed under the stars one night, on the way home from a ride to North Canterbury,' one recalls. Others slept nights in holding paddocks with their charges, and went on

their way again next morning, to be greeted by home-baked gifts and encouragement from farmers' wives en route and stunned stares from male motorists.[42]

As the men came home from war, however, women went back into more traditional roles, and fewer women were seen on the road independently droving stock.

While they were not officially land girls, June Leslie of South Canterbury and Ngaire Thomson from the Manawatu worked on the family farm during the war without any recognition: droving was just one aspect of their work. Ngaire drove stock to Feilding from the family farm out on Ridge Road; and June drove in Canterbury alone. Thoughts of a career went out the window for June in 1943 when, aged 16, she finished school and headed back to the family farm in South Canterbury to assist her father because of the labour shortages. Droving was part and parcel of her routine, moving both cattle and sheep – she took it all in her stride. She wasn't afraid to stand up for herself either, as older drovers sent messages via a passing motorist to get her mob off the road so as to avoid stock being boxed.

> I think probably the cars said, 'There's a girl down the road with sheep.' And they'd send a message down to me 'get into so and so road while I pass'. And I'd say to the person who gave me the message 'Where is the stock?' 'Oh way back so and so.' And I'd say 'Well they're closer to a road than I am, they can get in.' And I'd send a message back 'You get off the road for me.' And I felt so cheeky – but they didn't have to wait nearly as long as I would have waited.[43]

June Leslie brought a feminine twist to droving:

> I always took my knitting with me. You know I'd drape the reins over the horse's neck at times and just sit there and knit or I had a little bag with the wool in and just walk along the road and knit. It's just what I did, there was nothing interesting about that – I thought – I mean I used to get a lot done and it was quite good.[44]

For Betty Eggeling it was tatting that occupied her:

> I used to put in my pocket a ball of cotton and I had a tatting shuttle, and I used to tat away while I was following the cattle along, 'cos you're on your own, you haven't got anyone to talk to.[45]

Sometimes an entire family would go out on a drove. Bill Pullen, a master at conjuring the delights of droving through words, convinced his wife Gill in the mid-1970s that being on the road would offer fun and a 'relaxed lifestyle'.[46] Gill agreed, and they bought a truck and caravan and, with their young family – two little boys aged three months and two years – drove cattle from around the East Coast–Gisborne region to the Waikato and King Country through the winter months. Gill went out in the lead, driving the truck towing the caravan. The image of a baby and a toddler, Bill, an extra male worker plus Gill, all in a 22-foot Oxford caravan, might have been fine and romantic when the sun was shining and the nappies were drying on a farm fence, but it was quite another thing when days of rain saw the caravan converted into a laundry and all were crammed inside.

The professional drover

In the twentieth century you would be hard-pressed to find anyone working on the land, whether farmer or worker, who had not moved the odd mob of cattle or sheep between blocks, to saleyards, railheads or the freezing works. While amateur drovers abounded, in the nineteenth and into the twentieth century a growing professional core of men emerged who specialised in moving stock on the hoof. Census figures show that drover numbers increased from just 110 in 1891 to a peak of 976 in 1936. This number is probably not truly representative, as there are issues around occupational classification[47] and definition.[48] What is apparent from the figures, however, is that there was

Members of the Gisborne droving fraternity at Matawhero Ford in 1901. *Tairawhiti Museum, 421-9_WFC-1-4-C35, photo by William F. Crawford*

a process of specialisation from which, ultimately, a knowledge and skill base emerged. Those interviewed often spoke with a degree of veneration for these 'real drovers' from the earlier half of the twentieth century.

Those 'on the road'

The drovers introduced in the following pages are a cross-section of the men and women who might be found 'on the road' throughout the twentieth century.[49] Charlie Sundgren and Harold Traves fit the profile of full-time drovers from the first half of the twentieth century. Arthur McRae, who would perhaps more likely call himself a stockman–drover, worked seasonally on the road while also working with stock – cattle and sheep – on many stations in the North Island in the second half of the twentieth century. Ben Morunga from the Far North did a considerable amount of droving for Lands and Survey and also drove independently right up until 2007 in the Broadwood area of the Far North. Two sons of drovers, Jack Curtis and Bill Traves, represent those who were drawn into the world of droving through the work of their fathers. Two women from opposite ends of the country represent those staunch New Zealand women who worked as equals alongside men in what was a tough and uncompromising world.

Sunshine Charlie

One drover who immediately captures the imagination is Charlie Sundgren, or 'Sunshine Charlie' as he was known in the Manawatu. While it is easy to assume that all drovers must have been rough and ready and earthy by nature, it was certainly not always the case. There were some who were consummate gentlemen, clean-living and highly regarded in the community. Charlie Sundgren was one such gentleman. His obituary read: 'As a man of strong Christian faith, high principles and integrity' he won 'a place in the hearts of all who knew him'.[50] 'Is the billy boiling?' was a catch-phrase that anyone who knew Sunshine Charlie remembers.

A drover of both sheep and cattle, Charlie was the son of a Swedish basket-maker. Born in Petone in 1892, he suffered a bout of rheumatic fever at age 19 and, on the advice of a doctor, he left the family business and went into farm work in Taihape and then in the Wairarapa at Glendhu and Pahau Stations.

After the isolation of station life, Charlie bought 10 acres of land along Kimbolton Road in Feilding, and moved his wife and young family there in 1926. It was there that he started droving to pay off the land. Although he did purchase another 10 acres, his intention to farm was overtaken by the needs of his family and he kept on droving up to his death in 1964. Initially he drove for any of the stock agencies in Feilding, but he developed a working relationship with Wright Stephenson that continued unbroken for 35 years.

Charlie was known to drove as far north as Gisborne and as far south as the Wairarapa, but his usual haunts were in the Rongotea, Oroua Downs, Bainesse, Halcombe and Levin areas. He was a regular at the Feilding saleyards, where he was nimble of foot, avoiding connecting with angry stock in pens. Charlie always intended to 'deliver the stock in better condition than what it left the farm – which meant he drove it slowly and carefully and allowed them to feed on the way so that there was no stress . . .'[51] This ensured he always had work, even through the depression.

Charlie's daughters Kath and Bev have warm memories of their father and the little rituals created around his work.

> At the end of the day I remember Dad coming home and always toasting his feet beside the coal range. As children we were always taught to go bring him his slippers and take off his big heavy work boots and put them on the rack above the fire to dry.[52]

He would tell them stories of the events of the day, the traffic on the road, who he had seen and any mishaps. Charlie loved to sing, as Kath recalls:

> One of the things I remember about Dad was he loved to sing and from our sleep we would hear the gig rattling down the hill of the drive way and Dad would be singing 'It's a Long Way to Tipperary' or perhaps one of the gospel songs.[53]

His daughters confess that many of the neighbours heard him too. Bev laughs as she remembers hearing how the grocer, Mr Saunders, on the corner of Lytton Street and East Street gave her father a bag of blackballs, saying, 'Put them in your mouth at four o'clock in the morning when you're going past so you don't wake us up.'[54]

Harold and Bill Traves

Harold Traves was a contemporary of Charlie Sundgren. Like many drovers, Harold did not start out his working life with the intention to drove. Born in 1895 at St Andrews, south of Timaru, Harold's education was curtailed at the age of 10 or 11 when he became a ploughman driving a six-horse team for his father, who was a contractor.

World War One interrupted Harold's working life when he signed up at the age of 21. He was badly injured on the front in France and returned home. After convalescing, he worked for a couple of years ploughing at Otaio, then bought a 12-acre block at the Levels in Timaru. His son Bill is not sure how the transition from ploughman to drover eventuated, but he was sure the war injuries, which damaged both of Harold's arms, were responsible: strong arms were necessary for ploughing. Harold's main income came from droving – Bill recalls him being paid 15 shillings a day – and even through the depression years he had regular work.

As well as local droves, Harold brought sheep out in autumn from the Mackenzie Country, including stock for the Cameron and Grant families, mostly to farms that grew winter feed in and around Rangitata; and in spring he drove the stock back up to the stations. The drove took six or seven days. On an annual trip from Birchwood Station at the headwaters of the Ahuriri River, Harold and another drover moved cattle for eight or nine days through to Pleasant Point for sale. There was the odd trip as far afield as Dunedin, and one three-month trip taking sheep from the Tekapo sale as far as Blenheim. Work in and around the district kept him busy as well. Over time the long droves were replaced with just local work, particularly working for Borthwicks, a meat-processing company, picking up fat lambs from farms and loading them onto rail for the Belfast freezing works.

Bill believes his own droving days began as a four- or five-year-old in 1925 or 1926. Somewhere in the recesses of his mind he recalls leading a mob of sheep across Caroline Bay. The mob, which had been shipped from the Chatham Islands, had just been unloaded from the *Tees* at the Timaru wharf. 'It was dark, getting on, the evening, you know, and he sent me out in front with a hurricane lamp to walk along the front . . .'[55] The full route is hazy, but Bill thinks they went across Ashbury Park and headed through back streets to the abattoir.

At age seven, with an entrepreneurial bent, Bill grew and sold garlic and broccoli for pocket money. 'I had a box on the handlebars of my bike and would fill it up with broccoli and go around the district and sell to the residents.' The garlic was sold to the grocer for 15 shillings a sugar bag.[56] In the summer holidays he would regularly attend the hot and dusty Pleasant Point and Temuka sales with his father, loading sheep onto the rail wagons after the sales. At 17 or 18 he often spent Sundays helping his father with the Borthwicks work, loading fat lambs onto rail. 'Sometimes I'd go on the bike with a dog on a string and collect up one or two lots and meet up at a pre-arranged place.' Bill appreciated the extra money: he earned as much in one day's droving as he earned in a week as an apprentice builder–joiner.

Harold stopped droving sometime in the 1950s as trucking became more viable. Bill had no intention of carrying on in his father's footsteps; when he returned after World War Two his focus was on a career in building; he later went into business in Timaru as Traves and Ward.

As Charlie's and Harold's careers in droving were coming to a close, another North Island drover's career was building momentum.

Arthur McRae

Arthur McRae was an unlikely candidate for droving, yet he became a well-known and respected drover on the East Coast for about 50 years. He was born in Wellington in 1931, the son of a policeman. He and his parents and eight siblings moved often, following his father's work – to Levin, Akaroa, Christchurch, Palmerston North and, finally, to Wairoa. He worked on farms in the holidays, and when he left school aged 14 or 15 he pursued his love of rural work, beginning as a cowboy on a station in Wairoa. There was no wild west Stetson, chaps, spurs – not even a horse. A 'cowboy' in New Zealand was a lowly position: Arthur was required to milk the house cow, chop the wood and mow the lawn. Over time he learned new skills like crutching and drenching, and graduated to owning an 'old dog and starting to bring the killers in for killing'. He absorbed new skills working alongside older men, and finally gained a position as a shepherd. Part of shepherding in those days involved taking stock to the rail yards and loading the wagons – the genesis of his droving career.

Arthur and his brother purchased a scrub block that needed work to make it a viable farm. To supplement their farming income, he took on any local work available droving stock in to the Wairoa saleyards. Gradually, as is apt to happen, the droving work developed a life of its own, and Arthur took work that required travelling greater distances – Wairoa to Hawke's Bay, Gisborne to Hawke's Bay; as he recalls, 'it just grew'. With experience came a good reputation and plenty of work.

After the farm was sold in the 1960s, Arthur took on droving and mustering full-time: droving over the autumn and winter months, and the rest of the year mustering on various stations. He had a long association – for 20-plus years from the 1970s into the 1990s – with the Duncan family at Otairi Station near Hunterville in the Rangitikei. The six-week trek with 500–600 head of cattle went from Gisborne through Hawke's Bay, over the saddle road to Ashhurst, along to Colyton and into Feilding then on to Hunterville. For Arthur it became second nature. Partnering with other drovers where extra hands were required, he husbanded stock south, always aiming to arrive with the cattle in good condition.

He was a dedicated drover, but he laughs at the memory of jacking up grazing for one mob so he could pop over to South Africa to follow the All Blacks on the 1976 tour.

> I went over to South Africa and followed the All Blacks for a few weeks and then came back and carried on. And actually, I was doing it for Otairi – Chris Grace was the chap in charge there then – and when it came to pay us for those days . . . I said, 'Come on, I was away for some of that time.' And he said, 'That's alright, you were still in charge of the cattle and if anything had gone wrong you'd have bloody known about it.'[57]

In 2008 Arthur was still working with stock. As he says, 'I've been the only one droving round here for years now, there's no other drover . . . anyone wants to shift anything they get in touch with me.' The droves were local, however – around Wairoa and Gisborne; the days of the big droves were long over. There was increased traffic, and the council's attitude to stock on the road had changed: 'Oh, the rules and regulations . . .' With regret in his voice, Arthur philosophises, 'As I say, that's what they call progress – times change . . . It's just a pity . . . That's coming from a 77-year-old not a 27-year-old.' As he reflects back over his years on the road he's quick to say, 'But I thoroughly enjoyed it, yeah, thoroughly enjoyed it, made a lot of friends, and I got into a few arguments but usually got out of them [laughing with some chagrin], yeah.'[58]

Ben Morunga

Ben Morunga's story introduces the reader to one of many Maori drovers. It also typifies the transient nature of a drover's work.

A quietly spoken Maori man, Ben Morunga was born in the Hokianga in 1929 of Te Rarawa descent. He drove stock for over half a century in the Far North. Droving was by no means a full-time working occupation for Ben: in the beginning it was a necessary supplement to income, and it later became part of his everyday work when employed on Lands and Survey blocks.

Ben began his working life at around 18, tree felling in the bush at Herekino with his father Tom Morunga (known as Tom Clark), who, from his late teens, had worked with bullocks in the forests, mainly in Northland but also as far south as the Gisborne area. Ben talks of the bush life being hard, but he reckons, 'it taught us a lot . . . it taught you how to work and respect people and all that, eh . . . and mix with different people, you know . . .'[59] – lessons that no doubt stood him in good stead when he was on the road with stock.

His father bought 90 acres at Panguru and established a herd of 15 cows for milking – a good herd for those days. This along with money earned from bushfelling kept the family of 15 afloat. In Ben's early years, milking the cows before school was all part of the daily routine; and after his mother died when he was just 11, he helped look after his siblings. Breaking in horses and retraining other people's horses was also part and parcel of his early life.

As bush work dwindled, Ben took on more farming work. He gained experience working with stock, shepherding on farms in the area, particularly on Maori Affairs blocks.[60] Some years after the death of his mother, Ben's father retired to Auckland and Ben took over the farm. It was a time when many families were selling up and moving to the towns. Ben purchased land and built the milking herd up to 35 cows alongside about 15 head of dry stock. He still worked away from the farm during the off-season to help pay off the mortgage with Maori Affairs.

Droving became a part of his working life during this time. Initially it was 'chasing' cattle – short droves of only a couple of hours picking up stock for farmers around the Broadwood district. He was about 21 when he first did a trip of any distance. Later he drove boner dairy cows for Maori Affairs to the works at Moerewa, a five- to seven-day drove; the mobs got as big as 500, and Ben remembers the trouble he had getting extra drovers for the job. A memory of one hard-case inexperienced drover makes him laugh: 'About 10 o'clock in the morning he started asking, "How much further? . . . how much longer? . . . how much longer?"'[61] In the finish Ben, seeing that the mob was settled, moved up to the front of the mob so as not to hear his complaints.

To add to the mix, local stock agent Eric Carmen tracked Ben down and gave him casual droving work for Lands and Survey. For several seasons he drove mainly cattle further north to a block at Sweetwater in March/April: seasonal transfers of stock between blocks in the Far North and more southern blocks was the norm. The Lands and Survey work landed him a permanent position as a shepherd for the department at the end of the 1950s, and he was later promoted to manager: he was proud to be the first Maori manager on a Lands and Survey block north of the Auckland harbour bridge. Droving was still very much part of his work as a manager.

After he retired from Lands and Survey work and concentrated on the family farm, he kept carrying out local droving work for himself and his family and for others right up until 2007. Droving offered a cheaper alternative for small local droves: a mob of 100 head would take two big units [trucks] at $1500 per unit; whereas Ben charged about $100 a day.

When asked at the age of almost 79 if he would keep moving stock, Ben replied, 'If it's worthwhile, a worthwhile mob, I don't mind. I'll keep doing it . . . I like it . . . I like it.'[62]

Jack 'Boy' Curtis

Like Bill Traves, Jack Curtis, or 'Boy' as he was known locally, was roped in to droving by his stockman father. Unlike Bill, however, Jack loved working with stock and eventually bought his own farm, partly funded through his droving work.

Born in Gisborne in 1937, Boy grew up in the shadow of his father, stockman Jack Curtis – a hardworking stockman who worked stations around the East Coast–Bay of Plenty region. He was away a lot, mustering or droving, and Jack's young family all had to 'pull their weight' to help their mother. Boy recalls it being 'bloody tough'.

As a 10-year-old, Boy recalls going out on his little pinto pony with his father to bring in stock from the back country to the Gisborne sale; he remembers the days where 8000–9000 cattle were yarded for a sale. At the age of 12 he started working dogs at weekends; in 1950 he won a prize in a dog trial with an old dog called 'Old Joe'. By then, school took up two days a week and the rest of the time he was off shepherding.

With the knowledge shared by his father and the advice from old station staff, Boy was starting to develop his stock-handling skills, and by 16 he was taking any opportunity to work on the road. But Boy had aspirations:

Left: A young Boy Curtis stops for a photo just out of Opotiki. The mob of sheep he is droving continue on their way. *Jack Curtis Collection* Right: The weathered face of Manawatu drover Keith Sowerby suggests a lot of time spent in the open air. Keith drove for over 50 years in the district. *Julia Sowerby Collection*

'I loved the job, but I wanted to buy a farm.'[63] He watched others on the road drink away their wages, but not him. Droving and other stock work kept him busy – and he saved.

After he married, Boy leased a broken-down old farm. He survived more by hunting and doing the odd drove than by farming. Once he had a young family he let go of the droving:

> . . . but when you've got commitments . . . ah . . . you've got to think – now, is it worth it? Am I doing it for fun? Or do I need the dollars? But it's something, maybe – I don't know, maybe you'd say it's in my blood – but it's something I've always wanted to do.[64]

Boy purchased a farm in Whangamomona in Taranaki in 1981. He still did some droving occasionally, after this, but it never played the significant role it had in his younger years. Despite his droving days coming to an end years ago, when he talks about it Boy's eyes dance with the memories of the days on the road, and the insights and stories that he expounds on fill several hours.

Myrtle Pratt (née Cron)

Myrtle Pratt (1916–1981) was 'a legend in her lifetime'.[65] She grew up on the Landsborough run in South Westland. The granddaughter of Adam Cron – one of the original settlers of the Jacksons Bay settlement – Myrtle proved herself to be of true pioneering stock. A jack of all trades, she farmed a cattle run up the Haast River, cut fenceposts, and ran the small Haast airport.[66] A redhead who was 'known as a bit wild',[67] Myrtle had a reputation for a fiery temper; but she was held in high regard and was a bit of a legend in the area. Allan Barber, a stock agent for Wright Stephenson, recalled that 'she had an excellent knowledge of livestock . . . and she ran that big Landsborough area particularly well'.[68]

History books make mention of her droving cattle up to Whataroa; one account states that she did it single-handed.[69] Those who knew her tell other stories that reflect the strength of character that enabled her to get the job done. Her nephew Allan Cron remembers her fondly:

> Myrt was a great sort, she was my favourite aunty. She could swear like a bloody man, she could drink piss like a man, you know. She loved dogs, she loved horses. She made fools of the dogs, the dogs became so protective of her that you were lucky if you didn't get your bloody legs chewed off, you know. She loved cattle, she loved mustering. I mean she was dying of cancer and she was so determined

to come up that bloody river and muster up the Landsborough. And we waited right until June and she came up there and she was dead in October . . .

She loved the outdoor life, the mountains, the rivers, she just loved being there. She could build a barbed-wire bloody fence. She built a better stockyard. She could shoot deer. When she was younger she shot a helluva lot of deer and skinned them to help with her income.

When her father died – my grandfather – her father died in 1953, she took on the running of the estate . . . Myrt took on the running of that place, which was a bloody mean feat to do, to take on the rivers and everything else up here, but she took it on and run it successfully. And she drove cattle on her own up to Whataroa . . . She had three very good dogs – Ned, Spark and Dusky. And those dogs would just do about anything for her . . . Her dogs . . . knew all the nooks and crannies. She excelled, she really did.[70]

Tony Condon of Paringa laughs as he recalls droving cattle to Whataroa with Myrtle.

Myrtle was the first [woman] I ever heard use the 'f-word' . . . Halfway through the first day there was a big slip, Slippery Face it was called – we had a lot of trouble there. I finished up at the back with Myrtle cos I'd left my horse and I'd been on foot cos I was the youngest and I could travel, I suppose, a bit quicker on foot cos that particular place wasn't really horse country. So I ended up at the back . . . the cattle had got really stirred up. And we got up to the top of the track heading this way and things had sort of settled down. But apparently this bullock had got separated, he was a Cron bullock,[71] and he got separated from his mates and he was more looking for his mates than anything else. Anyway, by the time he got back to us I think he'd decided that Haast was his next stop. He's coming at us, horns like that [arm's length] . . . How am I going to stop this? I'm just not. I'm just not. Then the next minute Myrtle lets out a roar, 'Let the f—er go!' So alright then, I let him go.[72]

The South Westland drove was not for the fainthearted, and Myrtle's independence and take-no-prisoners approach to life probably held her in good stead when she was out with stock on the rugged route. A lasting memory of Myra Fulton's – another Okuru member – illustrates Myrtle's determination:

> She came through to Jacksons Bay with us for a dance. We had to walk behind Mt McLean and somebody wanted to carry her bike for her – she wouldn't take the truck, she biked. And someone wanted to carry her bike for her and she wouldn't hear of it – she was awfully independent. Put it over her shoulder and away she went. Well she got in to the Bay before we did. She was an amazing person.[73]

Phyllis Clements

Born in Whangarei in November 1926, Phyllis Clements grew up on the family farm at Matapouri–Woolleys Bay.

> One of the earliest memories I've got is Dad putting me on the front of the saddle on the horse and I'm not sure what age that would have been – it would have been three, I suppose . . . roughly round about then. And from there on I used to go with him on the farm, mainly at that age just shifting cattle round and that sort of thing.[74]

As a young girl she was required to help work inside the home, but it was the outdoors, horses and working with stock that inspired Phyllis. Her father was a cattle dealer and butcher, and he always had some stock that needed moving. Phyllis started by regularly moving stock into the killing paddock, then progressed to going on short droves with her father, out the front of the mob, shutting gates and watching for cars. She learned to ride aged about five by toddling around home on her pony; and by 10 or 12 she had her first working dog – a broken-in border collie called Sweep. By paying close attention to how others worked their dogs, she gained confidence in working her own dog. After leaving school at Standard Six she was involved in droving stock around Tutukaka and Ngunguru. She chuckles at the memory of being growled at when she did something wrong, but her lasting impression of this time is positive: 'Oh no, it was quite exciting, you had your pony and your dogs . . .'

When Phyllis moved to Kaitaia with her husband, Eric, sometime after her marriage, she had a reputation as a capable stockwoman and was offered a full-time job with Ken Lewis, working on his Northland properties. One job on the annual agenda was the Far North cattle drove that Ken ran for over 50 years. Phyllis has lost count of the number of times she drove for Ken on this Far North institution – she reckons seven or eight years, maybe even as many as 10. Her versatility made her an important member of a hardworking team. At the sales along the route she acted as Ken's scribe, her red notebook at the ready to take tallies of stock to drove. As head cook for the trip she worked with Ken prior to the drove to plan meals and accommodation; she has been known to prepare a hearty breakfast for the team on a primus on the back of her Valiant ute. Then there was her role as farrier for the drove; and, of course, as drover – where she was often found out the front of the mob negotiating traffic, hoping not to be skittled by oncoming vehicles. She was quick to wave traffic down, and with wry humour she comments that the drivers would be thinking, 'Oh, better do something – a pretty savage old duck up there, we better do something!'[75]

That 'savage old duck' was remembered by Ray Stevens, who had been on the drove with her, as 'certainly a very good horsewoman' who had 'good dogs' and who 'pulled her weight. It wasn't till you looked at her that you realised she was a different sex. But a very, very capable woman.'[76]

Despite the responsibility of the job, the hard work and the uncompromising environment – including being sandblasted on Ninety Mile Beach – as Phyllis reflects back on her years droving with Ken she is quick with her praise: 'Wouldn't have missed it for the world. It was really something I enjoyed, and still get enjoyment out of it – thinking about it, you know. It was a wonderful experience.'[77]

Droving, for many, was a means to an end rather than a career choice – a way of generating income when financially stretched, or for a deposit on a farm. And for the ring-ins – children and wives, or a local with time on their hands – payment was seldom given. There were those with a sense of adventure who wanted to get out and try things: 'Oh, I was just a gofer – go for this and go for that and "get that young fella", "chase that bloody bull" . . . I was just a general odd body, I wasn't a drover, I was just a school teacher who liked doing things.'[78]

For others it was a career, and one they loved and took great pride in. For Charlie Sundgren's daughters there was a sense of pride in what their father did; he was an 'honourable man doing an honourable job.'[79] Not all were saints – as Boy Curtis points out, there were some 'I wouldn't even trust with my dog'.[80] On the whole, while there were certainly characters among them, they were just regular people making sure a job got done and got done well.

Jack Richards drives cattle past the Wilsons' house at Puramahoi. *Maureen Papps Collection*

5. THE CRAFT OF DROVING

Day to day on the road

You've got all these things – you're sitting on a horse that's an individual, you've got dogs that are individuals and stock that are all individuals, and there's no rules or regulations as far as they're concerned, so therefore you have to have a skill to combine those.[1]

A typical day droving – was there such a thing? While there may have been a certain continuity of routine each day, whatever else happened throughout the day's drove was completely unpredictable. The dictates of weather, traffic, route and stock meant an endless miscellany of events that might occur on the road in any one day.

The day itself inevitably began early, at five or six in the morning; or, if it was the start of a new drove there might be a ride of many miles out to a farm or station to collect stock – in which case the drover would rattle his dags as early as two or three o'clock in the morning. Preparation was essential before departing on a long drove: horses had to be shod, the tucker box filled, gear checked and dog tucker organised. Droves closer to home required a little less planning, but there was still the early start. The morning ritual followed a pattern: getting up before it was light, boiling the billy, making breakfast, watering the horses and saddling them, stowing gear on the packhorse, letting the dogs off, storing dog chains on the saddle (just in case there was an emergency and the dogs needed tying up) – all before heading away.

Once they were on a drove, the stock needed stirring from their holding paddock early to avoid them marching around and making a mess of the paddock. The ritual morning count-out was always carried out as stock went out the gate heading for 'fresh pickings', unless the drover was happy that the holding paddock was secure.[2]

There was a knack to counting out stock, especially sheep. Cattle were counted out in ones or twos, but sheep were counted out in twos, threes or

Frank Lambert, a Gisborne drover, had an unusual method for ensuring a regular wake-up time unaided by an alarm clock. Every night he would have a full mug of cold water and nature would take its course, waking him at a regular hour.[3]

A Gisborne drover on the job around the East Coast. While droving along the coast looks idyllic, on one occasion there was warning of a possible tidal wave. Drovers Jack 'Shotgun' Jackson and Jimmy Eh! (Jimmy Atwell) were not fazed by such a warning, reportedly stating that they would 'tie their bedrolls to a tree and everything else could swim'.[4] *Tairawhiti Museum, 501-33, photo by Keith Wade*

fours. When a mob of several thousand sheep was on the move it was a big job, but the counter would get into a rhythm. With every hundred, a tally – a scratch on the gatepost or a stone slipped into a pocket – marked it. The gate was often only partly open to control the numbers heading through, though the 'real guns' could count with the gate wide open. If there was more than one drover, one would control the gate and another would tally.[5]

The number of drovers moving the mob would depend on how many head of stock there were. A drover could move up to 1000 head of sheep by themselves; more than that and they needed someone to assist them. With cattle, another drover was required once the mob was over 200 head – though whether the rules were adhered to religiously is another matter. Alex Cocks tells the story of droving as an apprentice with Len Bell in South Canterbury. After a few trips Alex was sent out on his own. One trip he had 1100 sheep, and Len told him, 'If anyone asks you how many you've got . . . tell them you don't know, so tell them to count them.'[6] Ian Mullooly, when

> Holding paddocks could be simply a corner of a farmer's paddock, a local domain or a school field. In the pioneering years, accommodation houses were required to provide a holding paddock for stock. In later years councils had stock reserves along stock routes, and there was usually a per-head charge for a night's rest there. Charges varied over the years: some farmers never charged, recognising a loose reciprocal arrangement: if they themselves were in need at some stage, help would be given.

asked why drovers always remembered the numbers in their mobs, said: 'Oh well, you had to count the bloody things that many times it sticks in your mind alright . . .'[7]

Once the mob was on the road, work began in earnest. People often assume that droving involved just following along behind the mob, with little effort required. Long-time North Island drovers Bill Pullen and Arthur McRae are quick to set the record straight on that. As Bill Pullen says:

> There's always something to do . . . People get the idea you're wandering along with your hands in your bloody pockets and you're doing nothing – you're always on the ball, you're always on the lookout. It's a very stressful job. And you're stressed all the time. Well I was anyway. But then again I never lost cattle . . .[8]

And Arthur McRae adds:

> Not many people could drove, they'd get bored. You know I had a young fella working with me and he was getting bored, and I told him he better toss it in. It's no good if you're getting bored; you're not doing your job – you know, you're not watching and listening.[9]

If the drover was working alone, the leading dog would be out the front and the drover would come along behind. If there was more than one drover, there would be one in the front ensuring a suitable pace, closing gates and guarding side roads and cuttings or banks until the lead of the mob had passed.

Each drover had different tricks to avoid trouble. Arthur McRae used to 'take baling twine and . . . tie it across open gateways and fences and . . . hang a bit of paper over it and they wouldn't go near it then you see.'[10] Bill Pullen, knowing the route, would go ahead and chain a dog to a fence where he knew there was a cutting or bank, to avoid stock falling off. Sam Jefferis talks of droving through the Waioeka Gorge on the road from Opotiki to Gisborne. You took nothing for granted, he says; you trusted nothing to chance – you were always checking.

> If you're walking on the side of the road through the gorge and there's a river there, you walk on the outside edge, you're looking and you might see, 'Oh something's gone down there,' and you might be two miles before you get the beast back up again. And those are unexplained skills you never learn out of textbooks – it just comes from experience and knowing what you are doing.[11]

Top: A large mob of sheep on the road near Atiamuri on the Waikato River in 1908. Bottom: Sheep are driven along the main road in the vicinity of Hanmer, North Canterbury, in 1930. *Sir George Grey Special Collections, Auckland Libraries, AWNS-19080402-3-2 and AWNS-19300423-44-2*

'Teaching' stock

Conventional droving wisdom said that it took about three days before a mob of cattle had settled into a routine. While it's difficult to imagine stock as willing students, drovers talk of teaching stock to come out of holding paddocks and feed. As Spencer Dillon puts it:

> You'd get cattle and they hadn't been well handled . . . you'd get them out and they wouldn't feed – only one thing to do: take 'em back and put 'em in the holding paddock till they got empty and then they'd feed and settle down. It usually took about three days to settle cattle down on the road.[12]

Ray Stevens recounts his experience of the stock's response to the routine:

> The first night or two she's reasonably hungry because she's lost a lot of the time, everything is strange to her . . . the second morning long before – well – 11 o'clock they're hungry and they've got their head down, third or fourth morning the moment as you open the bloody gate they're eating – they soon get into the rhythm: 'Well if we don't eat now we'll be hungry again tonight.'[13]

Even the most obstinate of stock could learn. Ian Mullooly remembers one mob – 'and boy oh boy were they wild' – they had come out of the backblocks of the East Coast at Te Kumi. Ian quickly decided, 'I'm going to root these buggers along (excuse my language), so I made them go and I did the 18 mile down to Waikura.'[14] Sensing this was still not enough to make them civil, he penned them in a stockyard overnight so in the morning they were ready to feed. With such treatment they soon settled down.

When the drover took the opportunity to stop to boil the billy, the stock would soon learn to quickly lie down and rest. When it was time to move on, the stock might be given a bit of a shove to get them up, and on they would go again. Jack Curtis admits, though, that while most of the cattle got pretty quiet on a drove, some didn't: 'the Waipawa bullocks didn't. You didn't stand them up, they bloody stood you up!'[15]

Drovers had their preferences when it came to what stock they moved. Often it was cattle, as they were more 'free-moving than sheep'.[16]

> I preferred working with cattle. You've got to keep driving sheep and you have to have the right dogs for it. The heat gets to them, specially on the tarseal. Oh Christ I hate sheep! They take a lot of controlling, specially when they've been driven into the sale and been in the yards

Drovers had to be willing students, too. There was much to learn from other drovers. Edgar Jones, an Amuri runholder in the 1860s, used to take stock to the West Coast to the diggings. On a return trip he came across the Elder brothers droving a mob of sheep over to the Coast. Edgar learnt a lot from watching how the brothers looked after their animals.

> We talked for a little time and they gave me all the news, but the sheep had not drawn on, so I offered to go on through the bush to the lead and give them a start. 'Oh no,' they said, 'don't do that, leave them, and they will move along directly': which they did after about 10 minutes. They told me that they put the dogs round the sheep as little as they possibly could. Up and down the rivers where there were grass flats, they let the sheep feed along at their own pace; they only kept behind so as to keep them in the right direction. At times the sheep would lie down to chew the cud, and they let them. At a river crossing they did as little dogging as possible. They left the main mob with a dog which was trained to hold them, and took a small cut of 20 or 30 sheep up to the river along a spit, held these sheep there quietly, but firmly, with themselves and their dogs, so that none of them could break back. They watched the sheep: when finding that they could not get back, one of them looked over the river – the only way he could go.
>
> Practice had taught them the right time to make their dogs bark, and themselves closed on the sheep. The sheep that was looking across, being pushed from behind by the others, went on and across. The main mob was then brought up quickly and they naturally followed the others. I learnt those lessons, and they were useful to me afterwards. If all drovers studied their flocks in this manner, it would be better for the sheep.[17]

Laurie McVicar recalls, as a lad at Totara Flat in the Grey Valley, 'My role models were farmers and drovers and stockmen. I wasn't interested in what bank managers and accountants were doing at all – or towns. I was an out and out country boy.'[18] Laurie was a keen student and hung on every word of his idol, Jack Esler, a drover-dealer from up the Tutaki at Murchison. Laurie describing him as 'an identity and a role model of mine and a fine old gentlemen – tough – tough as boots.' He remembered Jack's advice decades later:

> One of the things that Jack said to me was, he said, 'If you're handling touchy cattle, take plenty of tobacco with you – when things go wrong start rolling cigarettes instead of getting excited.' I've never smoked but

The nightly count into the holding paddock. Bill Pullen at Fisherman's Paddock, Bay of Plenty, in the mid-1970s. *Bill and Gill Pullen Collection*

Bill Pullen drives cattle across the Whakatane River, below the bridge at Taneatua. Crossing a river gave the cattle an opportunity to have a good drink. *Bill and Gill Pullen Collection*

Allan Crawford leads his mob from the Far North across Twin Bridges in 1984. Cattle could be unpredictable on bridges, and there was always the chance of some going over the side. *Shaun Reilly*

Bill Pullen's reliable companions at Fernhill on Maraekakaho Road in Hastings. With no sign of rain, Bill has the pack cover neatly folded and tied on top of all his gear. It was important to load the packhorse carefully to avoid any chafing, which could create sores on a horse's back. *Bill and Gill Pullen Collection*

Dick Gibbens with his loyal friend and hard worker, Bob. *Owen Gibbens Collection*

Left: This advertisement appeared in *New Zealand Farmer* in May 1918. An oilskin coat was considered an essential item for a drover, and would often be tied to the front of the saddle. *Hocken Collections, Uare Taoka o Hakena, University of Otago, S13-377c*

Below: An advertisement for the ubiquitous Cooper's sheep dip. The empty containers often doubled as the drover's tucker box. *Images from the History of Medicine, 101460207*

> I've always remembered it – if things go wrong just settle down and make out you're rolling cigarettes and it often comes right.[19]
>
> There was certainly much to learn, the old teaching the young. Spencer Dillon reckoned while the old drovers would 'tell you a bit', the best way to learn was to be 'working with them'[20] – although perhaps some habits learned were not so welcomed by the parents of budding drovers! 'And one of the things we did, of course you know, everyone – all these people smoked, so my brother and I, we used to smoke. I mean I knocked off smoking when I was 12, I suppose, I'd smoked up a few before that.'[21]

> all day and night with no grass. They try and get through every little gate open into people's gardens, over stone walls. You have to get some grass into them. Especially the Perendales. The Romneys aren't so bad. Those Cheviots – Christ, they're like goats.[22]

Robin Turner, droving in the King Country, recalled how careful you had to be with sheep.

> Sheep – if you're not careful when you're droving, and you push them too hard and there's a stream or a river or a creek somewhere, one of them will say, 'Oh blow this, I've had enough of this,' and will go down to the water's edge and stand there and defy you. If you go near 'em they'll jump in and drown perhaps. If you keep away from 'em, leave 'em alone, they'll come around and come out again. You have to be very careful.[23]

Jack Curtis, however, reckoned that sheep were more interesting as they were more cunning: 'they see a little bit of wool on the fence where a sheep has got through there about six or eight months ago and they know, "Hello, somebody's been here," and they'll run and jump through it.'[24]

Lial Bredin claimed that dairy cattle which were hand-reared were 'not really very much of a problem to handle'.[25] Weaners were 'as mad as snakes';[26] horses got 'lame and bitchy just like cattle';[27] 'sheep can be pretty piggy'[28] but 'easier on your dogs';[29] boner cows were 'just nightmares';[30] and moving lambs off their mothers was 'a very enlightening experience'.[31]

Laurie McVicar preferred droving cattle to sheep, but still maintained 'half a loaf is better than no bread: you're better droving sheep than not droving anything'.[32] For Ray Stevens, however, the only time sheep were okay was at 'night time with a few spuds on your plate'.[33]

One thing all drovers were unanimous on was that moving bulls was no picnic. It was a 'real tricky sort of job to do because they start fighting . . . I didn't like it at all . . . you've got to be really firm on them and then of course with bulls fighting, well they've got the right of way as far as I'm concerned . . . they're busy with what they're doing and not watching where you are, you don't mean a thing to them . . . you just move on, make sure traffic doesn't get tangled up . . .'[34] A job that should take three or four hours, with bulls could turn into an all-day affair as drovers battled to keep them under control as the animals fought for supremacy along the route.

Local Maori droving their pigs from Waipiro Bay to market in Gisborne in 1912. Prior to the introduction of trucking it was not uncommon to see pigs being moved on roads around the country. *Sir George Grey Special Collections, Auckland Libraries, AWNS-19121031-16-6*

Animal husbandry

It was a point of pride for drovers to arrive at their destination with the stock in good condition – or even better condition than when they left. Key to a successful drove was the drover's skill to husband their stock. Good drovers, Don Monk points out, 'let them along and feed . . . [they] took their time . . . might be two hours later but he still got their beer before tea, you know what I mean.'[35] On short droves of one or two days drovers could afford to push their stock at a faster pace and were able to cover greater distances in a day. Distances covered on a longer drove, however, ranged from as little as 5 miles (8 kilometres) a day to as much as 10 miles, depending on the terrain. It was important to ensure stock was moving at a steady pace, to prevent them 'knocking up' (going lame): at the right pace, they were able to spread out and feed as they went. As Bill Pullen succinctly puts it:

> . . . and of course you let them spread out for a reason; you let them spread out for number one: because you're not holding up any traffic. Number two: they're grazing better and they're not knocking their feet around, and really that's the key of the whole thing. Firstly, you've got to feed 'em and secondly, you've got to look after them. Well I mean if you don't feed 'em they get skinny, and if you don't look after them, well, they don't get there – simple.[36]

Feeding stock adequately was imperative – which was all well and good when you had the sole mob on the road, but often multiple mobs could be travelling along any one stock route, depending on the time of year, resulting in a shortage of feed if the mobs were running close behind each other.

Drovers had a variety of techniques to ensure stock got enough feed. One was to hire a farmer's paddock for a few days: while the stock were resting and feeding, the grass on the long acre was given a chance to get away again. Another option was to head down a side road and graze that out; or drovers might simply slow stock down to allow a greater distance between mobs.

All was fair in love and war, however, and there was definitely a 'first up, best dressed' policy. Drovers wanted their mob out the front and would go to some lengths to achieve this.

> Well there were about four or five mobs all heading south, and they were all sort of stationed around more or less from Tiniroto and Wairoa and all waiting for someone to move, and I walked through them. And they were all ready to go out and I pulled up in Wairoa at a paddock and I wondered, 'How the hell am I going to get in front and feed them and keep them in good order?' I went down to the railway and I priced

> it out and I worked it out that it would take three weeks to drove them to Eskdale . . . and I could truck 'em through for about 10 bob a head – well if I could charge out . . . it was a trade. . . . anyway I just walked into Wairoa – got them on the railway and jumped in front of everyone and was the first to get the good feed in Hawke's Bay.[37]

The satisfaction of getting one over on a rival was great.

> I had a big mob of old cows one time at Otamarakau and there was a joker . . . he did a drove . . . we both got to the holding paddock at the same time. Anyhow he let me put my cattle off the road first and then he put his cattle in behind me, he thought that would block me. He didn't know that I knew that there was another gate. So I got up and let my cattle out on the road with the leading dog. And he was still asleep, he was still in bed with his hitchhiker. Anyhow I sneaked out and the old cows sneaked past and away they went . . . Anyhow, as I was going past the caravan I went whoop across the back of the caravan. He jumped up, 'Whaaattt whaat?' 'See you.' 'Jesus! How did you get the mob out?' 'See you!'[38]

Stock were creatures of habit and those natural leaders out the front of the drove at the beginning of the drove remained the leaders throughout – consequently getting the best feed. In order to ensure the tail was sufficiently fed, the front of the mob might be driven up a side road with a lead dog, and the tail would be brought forward along the route to give them a chance to get fresh feed. In due time, after about half an hour or so, the leaders would be brought back into the mob and would eventually work themselves up to the front again.[39]

Injuries to stock did occur, even when the drover took care to move them slowly. Jack Curtis remembers seeing some 'very sorry cattle' when care had not been taken and they had developed sore feet:

> they walk pigeon-toe . . . when they walk they sort of scuff the inside of their toes and you'll see them and they'll bleed – they'll bleed like hell too. And what they've done is they're wearing one side of their hoof off.[40]

Bill Pullen explains what's needed:

> You're not hammering at them all of the time. We used to have competitions to see who could be the quietest the longest. You know

Sheep on the move in the Waikato District in 1932. One of the trials of being on the road was the dust generated by unsealed roads. *Sir George Grey Special Collections, Auckland Libraries, AWNS-19321130-50-3*

> going through places like the Waioeka Gorge and that was really hard cos it was rocky and the side of the road, cos it's all rock country and the roads are very hard – if you're constantly hooling away at them they're jumping off banks and they jump off on the road and they bruise the hell out of their feet and they get stone bruises. And that's why a lot of guys would end up with a lot of lame cattle. You can't do that – you're better to go do it yourself.[41]

Bad weather was a curse, too: rain softened the stock's feet, and when cattle moved on stones they would inevitably end up footsore – 'and then the next thing you'd have a lame beast or something and you didn't want that'.[42] Unpredictable weather could also lead to disaster.

> I took a mob of sheep to Oxford one time and I had a leading dog at that stage. It was a foggy morning and I had the sheep at . . . at this gravel pit. And I thought, 'Well I better not let them out while the fog's down.' So eventually the fog lifted about nineish . . . about nine – half past. I thought, 'Well that's it,' and let these sheep out. And I hadn't gone a mile and the fog came down again and just blacked everything out. And I put this leading dog out the front – a little black dog he was – and from Oxford came the Midland bus and he had a couple of prams on the front, on the hooks on the front of the bus. And he never stood a show . . . he just ploughed straight in. I think it was 26 or 28 that we finished up with – we cut throats, and broken legs, one we picked up a piece of skin about a foot across or 15 inches across, that was one that went under the dual wheels at the back. And anyway . . . he hit

the dog – the dog went under the bus and out the side and headed for home – nine miles to Woodend. He was there when I got home . . . That was that. Since that day I've always carried a knife on my belt. I thought, 'Well you never know when you're going to need it.'[43]

You might imagine the stockwhip to be an integral part of the droving picture. Those interviewed, however, said they used the whip judiciously or not at all. After all, as Laurie McVicar points out, 'the farmer from the time a beast is born to the time it's sold to the butcher, he's trying to put condition on it, so one thing he doesn't want is someone running round chasing it'.[44] Although those working with run stock might well find the use of the whip necessary, particularly when drafting on horseback in the yards: 'but in those days you wouldn't have gone on foot in the cattleyard – something would take to you, you'd get killed'.[45] Karl Jones remembers, as a boy in the 1920s tucked away in the backblocks of Karamea, Dick McKay and Peter Anderson coming in and taking out cattle that were more used to the bush than man.

In the early years of settlement the whip may well have been used excessively, but later drovers learned to be judicious. Many drovers preferred to use supplejack rather than carrying an actual stockwhip. *From* Crusts: A Settler's Fare Due South *by Laurence J. Kennaway, Capper Press, Christchurch, 1970*

> I'm not quite sure which comes first, but you had the man, the horse, the dog and the stockwhip in that order, I think, was about the way they went. Because if you were on a horse and trying to turn an animal, well you couldn't do it better than by galloping like mad and using the stockwhip and cracking it in front of its nose. Oh, it was a real skill – about a 14-foot leather stockwhip, and when it cracked it really cracked, you could hear it all over the place . . .[46]

For young lads there was a sense of pride in being out with stock with their whip. It took practice to be a dab hand – and some took it to the extreme:

> My brother and I used to practise . . . taking things out of one another's hands . . . And I think we actually, finally, we got to the stage of taking a stick out of our mouth as well . . .[47]

Drovers tried to avoid having to leave cattle behind, but inevitably even the best had times where stock needed to be railed or later trucked, or arrangements made to leave them in a farmer's paddock until the drover was next coming through. Some form of payment was mutually agreed on by drover and farmer. Arthur McRae, when he was working for Duncans' Otairi Station, would assess all the stock purchased in Gisborne district by his boss and veto any stock that he considered not strong enough to make the long drove down through Hawke's Bay, Manawatu and up to the Rangitikei. Such culling increased his chances of success.

Bill and Gill Pullen used their ingenuity on one of their droves with cattle through the Matahina. One beast was quite sick and they needed to get it to a paddock further on so it could be rested and brought back into the mob as they passed. Their handy little Mini ute with a canvas canopy with windows on the side became the transport. Gill laughed at the memory of the service-station attendant taking a second look as he filled up their tank: 'He couldn't believe his eyes.' Gill comments the axle was 'down a little'.[48]

Sheep had to be shepherded well, too; again, the drover aimed to keep them at a steady pace. Sheep were prone to tonguing and lameness if pushed too hard – especially animals that were heavy with wool. A sheep that was knocked up on a long drove was most likely to end up in the camp oven, and the excess used for dog tucker. As Betty Eggeling commented, there were 'no poncey dog biscuits back then'.[49]

Local knowledge

As well as mastering skills in stock care, drovers developed a wealth of knowledge about the districts they travelled through. With experience they became familiar with the designated stock routes and key features to watch for along the way: gorges and their difficulties, cuttings, when service cars and tourist buses would be driving through and where to layby in order to let them pass. They knew the pitfalls in and around towns, where extra assistance was needed in order to come through unscathed; and where there were good open areas to stop for a break and maybe boil the billy. Those who were not familiar with the route would often reconnoitre before the drove. Ray Stevens, droving dairy herds from the Waikato to Whakatane in the 1970s, would travel with the owner along the route, taking notes of any danger spots, railway crossings, cuttings and the like. He would also take the opportunity to organise holding paddocks along the route. And each day he would brief those assisting him as to what to look out for on their route.

Cattle are driven along a treacherous stretch of beach in Taranaki, north of White Cliffs. As well as keeping an eye on the tide, drovers had to watch for rocks. Carol MacKenzie recalls: '. . . I was always terrified going over stones, cos I thought if my lovely [horse] Goldie broke a leg what would I do? I'd be in the shit proper.'[50] *Sir George Grey Special Collections, Auckland Libraries, AWNS-19340815-40-1*

The Te Horo tunnel provided access to and from the beach at the northern end of White Cliffs in Taranaki. It continued to be used by drovers up until the 1960s. *Puke Ariki, New Plymouth, PHO2007-246*

Drovers heading up to Ross knew to watch out for 'windmill corner', where cattle could be spooked by the sound made by a little windmill constructed out of the tin from a tobacco case. West Coaster Jimmy Ferguson remembered it well:

> There used to be a blowhole between the Big Waitaha and Pukekura. Billy Holley used to be the roadman there, years ago, and he rigged up a windmill, and had the wind coming out through the neck of a beer bottle. This ran the windmill, and it worked for years. . . . The windmill didn't spin all the time, but most days it did. They used to call that the 'windmill corner'.[51]

Droving to Arrowtown around Lake Wakatipu in the 1890s, drovers had to watch out for a right-hand bend at Rat Point:

> Arriving at Rat Point, there is a right-angled bend in the track, and, bush cattle not being educated in trigonometry beyond the virtue of straight lines, the whole mob went over the track into Lake Wakatipu, a clear descent of about 1000ft. Strange but true, only a few of the cattle got cut and bruised in their aerial journey, and the whole lot were swimming about the lake quite contentedly until safely landed.[52]

Where necessary – perhaps because of lack of feed, or changes in council regulations – inventive sorts reconnoitred new routes. When Bill Pullen needed a route with better feed he found he could travel from Te Teko to the Matahina Dam through the forest and right through to Reporoa, and then follow the Waikato River. This more southern route provided 'great feed' for his cattle and avoided the mobs travelling the popular route through Rotorua or over the Kaimais.[53]

Experienced drovers knew just the right way to deal with tricky stock. Laurie McVicar, eager to learn, always listened to Jack Esler:

> Jack Esler told me about the time in the Inangahua area where he was at a sale and these cattle come in and different people said, 'I bought that beast last year,' and someone else said, 'I bought him the year before.' And what had happened is that they had droven away and the beast had got off the road down a track that he knew into the river, and they had to go on with what they had. So Jack said, 'Nobody wanted him so I bought him with others.' And I said, 'How did you keep him out of the river?' and he said, 'I didn't. I put the lot in the river and took him away.' And once you've got cattle off their home

ground they're lost, and that sometimes is only through a wire fence – it doesn't take much to do it. But he put them in the river, kept them there, all of them, until they're off their home ground then brought them out on the road and that was the finish of the cattle . . . tricks of the trade.[54]

Daily progress

With all this in mind the drover and their charges slowly meandered along their route. As the day progressed and as opportunity allowed, a smoko break or lunch stop was taken, but only where safe – a nice straight stretch where stock could be seen feeding, or a layby, a big open area where stock could rest. Then onward they would go, with holding paddock organised for the night's stopover. There might be a brief pause at a rural letterbox to read the newspaper while stock meandered past – of course, the paper was always returned.[55]

The day might run smoothly, but there was no room for complacency. David Stroud, who was droving in the Manawatu in the 1970s, maintained, 'When everything is going really well, that's the time when you need to be really awake . . .' In a split second a peaceful scene of drover passing the time of day, 'taking the Government apart as you do' with a local cockie while the stock graze and slowly wander on, can turn to bedlam. Such a bucolic scene played out on one drove for David. A panicked motorist flipping his car caused David's mob to bolt, flattening about 'five chain of fence' and almost mixing with other stock. After preventing the shocked driver from trying to re-enter his car to regain his lost slipper as petrol poured out of the vehicle, David then had to race off and retrieve his wayward mob.[56] All in a day's work.

There was always the possibility of stock wandering into unfenced bush or tea-tree: time was lost sending the dogs in to chase them out. Or a stray beast might join the mob and need drafting out. Stock might take a fancy to stock on passing farms, and a meet-and-greet session would require further drafting. Where mobs were sharing the route, there was always the chance of a box-up – the mixing of mobs was something always to be avoided. 'Holy hell! – you go and get 1000 wethers out of 3000 ewes – would be a hell of a job!'[57] At times the drove might wend its way through a farm, in which case extra vigilance was needed, and a track would be cleared by working a huntaway to hunt everything away from the mob so no stock got mixed. This was one occasion the drover would punch his stock on, to move through the farm as quickly as possible – you did not want to be accused of grazing your stock on someone else's land.[58]

All the time along the road a drover, as opposed to a 'hooler', showed consideration for those living along the stock route. 'You've got to look after all the people you know – you can't just run roughshod over people cos they soon get pissed off . . . That's where all the complaints come from and next thing you've got all these problems.'[59] The reward for consideration: a hot cup of tea and some fresh scones.

> . . . women would come out with a cup of tea for you cos they knew damn well that the stock never got on their lawn outside there – you'd just ride up there and you know she had a bit of fancy stuff along the road edge, so you just stopped there and rolled a smoke while the cattle drifted by and she'd come out with some scones and a cup of tea and you'd say 'thank you very much' and you're away. Cos she's rapt. But other guys would go along and trample everything and knock over mailboxes and generally give drovers a bad name.[60]

The drovers, having faced the challenges and averted disasters (more of which are recounted in the next chapter), wanted to be counting stock into the night's holding paddock before dark. Arthur McRae emphasises that they were just 'holding' paddocks.

> They were just big enough to hold a mob of cattle, that's all you want. And the smaller the area the better for me, because you didn't want cattle wandering around – if you had plenty of room they'd march. You wanted them to get off their feet and lie down. That's what you want. It didn't matter if you were going a few days or a week, but when you were doing long haulage and you're on the road for months it's their feet – all you're worried about is their feet, nothing else – get them off their feet. You know being a few weeks on the road, that sort of thing, they settle down and you'd get them in a paddock and they'd just lie down pretty much for a day.[61]

Drovers using their wits

Stories abound of drovers who used their quick tongue and their wits to extricate themselves from trouble. Allan Barber, an old stock and station agent, remembered Otago drover David Donaldson's quick response when challenged about having no one in the lead of his mob.

> . . . he was driving a mob of cattle through Green Island, and you were supposed to have someone out the front as well as the back, and he

didn't have anyone out the front and there was a traffic officer there hauled him up and said, 'Hey, look here, where's your man up the front?' And he said, 'He's bloody well where you should be,' he says, 'he's at the front.' This was when the war was on. And he carried on.[62]

Jack Curtis recalls droving over the Kaimais with Ned Macky, a 'big, tall, placid Maori fella', when a car sped into the mob. 'The driver leapt out of his car: "You fellas are supposed to have a leading dog in the lead." And Old Ned Macky says, "Why, do you want to run that poor bugger over too?" he said.'[63]

Bill Allen was another quick thinker:

> Old Bill was driving along through Riwaka one day and there was an open gate into a hop garden and a ruddy steer poked off and went into the hop garden. And old Bill sent his dogs in after him, about three dogs on this steer going up and down the rows of hops. And the owner of the hop garden heard this racket and finally came to see what on earth was going on. Here was this steer careering along, madly pursued by three dogs, and just as he got there it burst out onto the road with hop vines tangled in its horns and so on. And so Old Bill rushed up to him, he was on horseback, he rushed up and he said, 'Leaving your gates open, knocking my cattle about, I'll sue you, I'll sue you, I'll sue you.' And he rode off before the bloke could see the smile on his face.[64]

Bruce Ferguson remembered Jack Flower's ways with words:

> As I say you could write a book about old Jack Flower, and he was an unusual man. He was doing a bit of butchering, the Onekaka iron works was working at the time, quite a lot of men employed there. And Mrs Hesketh ran a boarding house, and old Jack used to supply her with meat. He goes along this day and [she says], 'Oh Mr Flower that was a long skinny leg you left me last time.' 'Was it, Mrs Hesketh?' he said. 'Well the only thing I can think of,' he said, 'it would be, I've been killing some sheep that Sidney runs up on the mountains.' And at that stage Sid Flower ran those sheep from Lead Hills to Mount Hardy and Para Peak and so on, he ran about 1000 sheep up on there. He said, 'It's so steep up there, Mrs Hesketh,' he said, 'the sheep have one long leg and one short leg so they can go round the hillsides.' He said, 'You must have got the long leg.'[65]

Any drover worth their weight thought on their feet – or on their horse. From all accounts, Ken Lewis had the gift not only of conning farmers into providing holding paddocks at the drop of a hat, but also of extricating himself from trouble; even from the arm of the law. Mo Moses recounts an altercation with a landowner whose front garden had been damaged by stock. When the landowner kicked one of Ken's dogs, Ken hit the man with his stick. When the police questioned him, Ken quickly explained the event as he saw it: 'No, he jumped at me and he frightened my horse and the horse swung around and the stick caught him . . .'[66]

Quick thinking saved another drover from a beating. When Ike Grace, droving in the Gisborne region, found himself being sized up by an irate farmer he was quick to call on his mate Bob Tapune:

> . . . a hell of a nice Maori, and he was built like a bull. He'd been a very good wrestler in his day. And he had this drover with him, called Ike Grace, who was a little thin fella . . . And at Okoroire they used to camp in the summertime with the sheep in the school paddock – you know, the football field – and they liked the school because you had somewhere to have the shelter shed to stay in and everything. And it was Christmas Eve, and anyway there was a hole in the fence and the wethers were getting out into the crop next door, and the owner had been down celebrating and came driving . . . as he came driving home he saw the sheep getting in. So he pulled in and saw Ike Grace there and complained and wanted to fight him. Ike Grace said, 'I don't want to fight you, but if you hold on a minute,' he said, 'my mate quite likes a fight.' And when he came round the corner of the school – the bloke decided that fighting wasn't that good. And he took them home to dinner.[67]

There is a story told of Jack Condon not being quite so quick – much to his chagrin. Jack was taking cattle to Arahura just out of Hokitika. The trip from Mahitahi took some two weeks, and by the time he got there the price had dropped. Dissatisfied with the price, he refused to sell them.

> There was a dealer there that went over to Jack and said to him, 'I'll give you 15 pounds less 10 percent for those cattle.' Jack didn't want to let on that he didn't know what the 10 percent thing meant so he said, 'I'll think about it for a while and get back to you.' So he went over to the bar, over to the hotel. He asked the barmaid. 'If I offered you 15 pounds less 10 percent what would you take off?' She said, 'Everything including my socks.'[68]

End-of-day routine

Counting in would be done as the stock wandered into the holding paddock. Although drovers kept a careful eye on stock some did go missing, and there was nothing for it but to retrace the day's route to scout for the errant stock. If the count was out by just one, the drover had a bit of a dilemma:

> You'd count in at night – it was quite tricky at times, cos you're counting your mob in at night. And when you've got a big mob, like five, six, 700 cattle and you count them in at night and you get up to 699, 700 sort of thing and you might be 701. Well are you going to count them again! It's a hell of a job – you've got to put them all back out on the road and it knocks them around, and holds up traffic and God knows what – so you're not. So you think you must have made a mistake. So then you go on to the next paddock, well then you're five mile down the road. So you count them again and it might be 700, it could be 701 but if it is 701, which one is it? When you've got a mob brought out of a saleyards – if a mob's been brought from the farm and all one earmark it's okay, but [if] they've been brought out of the yard – it could be 20 different lines of cattle there, with 20 different bloody earmarks – you've got to go through every mark and look and check it out to find that particular beast or else just cut one out. Who are you going to give it to? Whose is it? Could be from the mob ahead of you . . .[69]

If there were three mobs on the road in a 5-mile stretch, numbering some 2000 head of cattle, the decision was more than a little tricky.

> You sort of get, if you pick something and you don't know where it come from and whose it is you don't sort of advertise in the Auckland *Herald* that you've got a beast – you know what I mean. You stick with it. Cos you're likely tomorrow to have one fall over the bank and break its bloody neck so you've got a replacement. You know, to be honest. That's the way things were.[70]

Once stock were paddocked, the nightly routine of taking care of horses, feeding dogs and preparing dinner began. Of course, if the drover was staying at a hotel or private home, they enjoyed the pleasure of a good dinner and the luxury of cleaning up in a bathroom. A beverage or two might be enjoyed, too, before retiring – and then the pattern was repeated the following day.

Enjoying a beverage

Some drovers were teetotal, but there were many more who felt the need for 'refreshment' after a long day droving, and the pub was a focus for many. Don Monk remembers when drink-driving rules were somewhat lax and 'so much booze was drunk in those days'. After a day stock-buying with dealer Bill Parkinson and his manager Les Chaney in North Canterbury, they would stop off for a drink on the way home.

> We'd call in on the way home, we'd stop at the Waipara pub and have a few there, we'd come back to Amberley and so-and-so might be there so we had a few beers there, then Leithfield, then Woodend and then they'd go to the club[71] here and I'd go home for my tea. I'd be that full of beer . . .[72]

Despite the ill effects of the drinking, the next day Don would be out early collecting the stock bought the previous day.

For Harry Frank the local was Rangiriri, and it took ingenuity to get there of an evening.

> I remember this chap – he was working on the rail. And he and I would get on the old hand-jigger and go down to the Rangiriri pub – leave the jigger off on the side, go down and come back and he'd get on the phone and ring up and make sure there was no train coming and so we'd pull it back.[73]

There are stories, too, of opportunities for liquid refreshment being seized before a day's work had even been completed. Neil Matthews from the Far North confesses:

> One trip I remember I was bringing – I never had time to do them straight after the sale, so I left them at the yards and got up early in the morning and picked them up, oh it wasn't too early, about half past eight. I was heading this way and a fella pulled up alongside me in a car and wanted to know if I wanted a beer. And I said, oh well it was a bit early in the day but you know I always had time for a beer. Mightn't get offered one next time if you refuse that one. So we had one beer and then we had another one and then a stock agent pulled up and he had a couple in his car so we drank them. Then someone else turned up. Well in the finish we were there all day. And Colin May lived a bit further down the road, and he'd seen these cattle and he'd gone out and put them in his paddock and come up to see what the

The Opportunist Drover

Brian Philps had done a bit of local droving from the age of 14, but no long droves. While he was working at Borthwicks an opportunity arose to work a long drove, and he jumped at the chance. Over 50 years later he could still recall the trip in detail. While he's not one for superlatives, the sense of pleasure that Brian gained from the experience is evident and reflects the desire for some to be on the road.

> It started off by someone coming over from the main office and saying that Ken Duncan had a mob of cattle up in the Hawke's Bay which he wanted brought home and . . . wanted two drovers. And I said straight away that I was a starter if I could get someone to come with me that had a leading dog, which was pretty important – a leading dog. And Rex Signal, he said, well he'd be a starter, so we were away. And from then on we had to get our gear ready. I had to buy a dog; I'd sold a dog to a mate of mine and I had to buy it back because I needed another dog. So we would have had four dogs each, and we arranged a cattle wagon at Ashhurst and we rode over to Ashhurst with everything tied to our saddles, with a big sack of dog tucker. Luckily when we got there the railway had supplied a T-wagon[74] . . . so we . . . put our dogs and gear in one end and our horses in the other end and then we got in the guard's van and had the slowest trip from memory in the train to Waipawa. And it was all exciting.
>
> And we got up there and unloaded our gear and put our horses into a paddock, tied up our dogs and went along to the pub and got a bed. Nothing was arranged, it was all 'she'll be right' stuff . . . And then we gathered our gear next day and headed out to get these cattle. I don't remember an agent, but there must have been an agent somewhere that was keeping in touch. So we went out and we hung our dog tucker in a tree in a farmer's paddock – didn't bother to ask him, just hung it there and away we went. Out across the Tukituk [Tukituki] and I don't know how far it was but it was quite a step. And we got out there and the station shepherds had the cattle on the road so we counted them – from memory – thought there was 240 – and took charge of them and away we went.
>
> We asked them, 'Where do we get a paddock for these cattle tonight?' and they told us – a cockie down the road. And I think they set us up, because when we got there and tried to explain we found the man, the cockie, was an alcoholic and he was feeling no pain. We were in big trouble and Rex had a very nasty way of laughing – when he shouldn't laugh he'd burst out laughing. So we were trying to keep a straight face and talk to this bloke.

Anyway the head shepherd came along and saved the day for us – we got a paddock and he took us over to his house and we stayed with him that night.

It took us two or three days – I think it may have been three days – to get to Waipawa. We had trouble on the Tukituk bridge: from memory we couldn't get them started, once we got them started they started to run and the bridge rattled and the more it rattled the faster they ran. I remember Rex put a dog out across the riverbed to catch the lead and it was a great run. And he just whistled him on and he had three stream beds to get across . . . and he just whistled him on and he finally caught the lead . . . Mmm, just a marvellous run. And so we got to Waipawa and stayed in the pub again, and on the Sunday, it was a Sunday, we took them across the river below the town bridge – which looked quite spectacular I thought – and on our way to Waipukurau. And I always remember that we were crossing the overhead bridge on the railway and a minister came along trying to get to church to run the church [service] so we had to help him through the mob so he wouldn't be late for church. And then down to Waipuk, a little bit of trouble on the bridge from memory – a lot of kids about on the Sunday – but we got them there, and then bowled along to the Tavistock hotel and got a room there. No real drama.

And we used to try and get them out by about eight o'clock in the morning . . . and I think the hotels must have cut us a lunch. I don't remember much – a long time ago – this was 1950 from memory. And away we went, and so to Takapau. I always remember them going up the sandhill, a big hill that's not used now, just let them drift – a great sight going up there. Down to Takapau, stayed at the pub there – nothing booked, just bowled in and get a room. And then down to Norsewood, and I remember it was blowing a gale there but we got them into a paddock – stayed at the pub again. Got more dog tucker. Dog tucker was a bit of a worry, there was no dog biscuits in those days and so we were always struggling to get enough dog tucker.

And we must have gone through the town – cos the road didn't go through like it does now – we must have gone through the town with these cattle. Then down to Matamau the next day. Stayed with Townie Beddingfield – he had a paddock and stayed at his house. And then we went in towards the hills from Matamau, turned off the main road – been on the main road – went off the main road from Matamau, and I always remember there was a big reserve and I don't know how but our cattle were in the reserve and there was lots of good grass and we were letting them have a good feed. From memory there was a quite a big gully in it, and I saw a car turn up the other side, they were so far away I couldn't see who it was, but it was a very clear morning and I could hear someone call my dog by name, so I knew it was someone that knew us. And then they saw the sign, the 'no stock allowed' sign, and they started to wave their arms and say 'get out of it', you know, 'we'll have the ranger on you', but I knew it was mates of ours. And it was the boss and some of the boys come up with some dog tucker for us and to have a day out. Mr Borthwick was paying for it.

So then it was we stayed with various people that had the holding paddocks. We had to pay paddocking for the cattle, which was probably about – well I'm not quite sure – may have been a shilling a head – no – probably about sixpence a head – probably paying about six pound a day. So we'd had to take money with us, Rex and I had put money into a kitty, we had to pay all the paddocking which we would get back afterwards.

Then we finally got down to Woodville and Norm Signal came and took us home; from then on we just stayed at home [each night as they were so close to home]. I remember when we came over the Saddle Road we were the seventh mob to come over that day. We had the biggest mob so we came last – there were seven mobs. So perhaps there'd been a drought in the Manawatu or something and they'd taken cattle over. We had to punch – quite a bit of dog work – got them up there and then down the other side. And I remember a great sight of them all strung out down there all walking single file in the water table – to get off the road – they were getting a bit lame, you see, on the metal, it was metal roads in those days – to Ashhurst and we gave them a spell the next day. We just fed them out on the road because their feet were a bit sore. Then we're over to the works cos the paddocks were cheap over there – we didn't have to pay paddocking. And then the next day was the final day down below Longburn to deliver. The manager rode through and I'd say he saw every beast – he rode quietly through them. And then he gave us a pat on the back and said he'd never seen cattle come over in better order. We'd had plenty of time – we fed them all the time. They were for Ken Duncan, and Ken Duncan must have been pretty happy because he didn't even bother taking the tax off – he just paid everything.

So that was our big trip. Loved every minute of it. Sorry when it finished. I was often tempted to take [droving] on, but it was just as well I didn't because droving was on its way out – lorries were getting more numerous . . . Oh they were good days . . . We just sat on our horses and walked along and yakked, and now and again we'd go up to the lead and see if the lead was alright. We had a leading dog out all the time. But the cattle had been on the road and they just fed along no trouble. We enjoyed it. We didn't have any problems.[75]

hell was going on, so he had to have a beer too. Yeah it was about half past five or six o'clock when we finished up there. So the cattle didn't really get delivered till the next morning.[76]

Gill Pullen remembers strife over pub visits when she was off droving with her husband, Bill. Stopping at a campground gave Gill the opportunity to catch up with laundry and be a little closer to civilisation:

> . . . but of course, doing that – stopping at the camping ground – was also handy to a pub, which wasn't quite so good either, because before coming home for dinner they'd sort of veer in and into the pub and it could be sort of – you're thinking, 'Gosh those cattle must be walking late tonight.' You know, sort of eight o'clock at night when it was pitch-black and that, and then in they'd roll quite merry . . .[77]

When they stopped at a friend's home in Ruatoria, where there was feed for the stock for a week, Gill was once again caught unawares. The men were disappearing, 'checking on cattle' during the day and coming home late for meals. By chance she discovered on a trip into town to pick up bread with her young boys – 'and here's the van parked outside the hotel at 11 o'clock in the morning'. A truce was eventually negotiated, but not before there was considerable dissension in the ranks.

Open camping

Drovers around the East Coast and through to the Waikato, in particular, might have to revert to open camping in more remote areas where there were no holding paddocks or accommodation. A suitable no-exit road or layby would suffice, somewhere they could hem the stock in and use the dogs to hold the stock together. When travelling with trucks, in the later years, scrim or even a portable electric fence unit could be used to make a holding pen of sorts and the stock would settle. The drover always had to remain alert, though, as a weather change or a fright could see the stock scatter. The dictates of the Waioeka Gorge, for example, meant four or so nights open camping. Sam Jefferis speaks from experience when he says, 'It's no fun when a truck pulls up and tells you that there's cattle five mile down the road . . .'[78]

While you wouldn't ordinarily think of cattle as being particularly devious, Gill Pullen laughs at the memory of one open camping site through the Matahina where they had 'circled the wagons', so to speak, with trucks and caravan. In the sure knowledge that the cattle were secure, they slept soundly.

> Well we woke up in the morning, we never heard a thing in the night, and when we woke up in the morning there would have only been a quarter of the cattle . . . they had stepped out between the caravan and the truck over the towbar all night long, one after the other, and got out, and they were miles away . . . Gone two stages in one night. All night long they must have been doing this and not one of them touched it so we could have heard them.[79]

The saving grace was that the cattle had conveniently travelled in the right direction.

Drovers could be on the road with a mob for a day, a couple of days, a week, a month or several months – one as long as 150 days.[80] If horses had odometers the miles clocked up by a drover in a single year would be significant; and with those miles came experience and knowledge and the chance of greater success. 'My theory was the first day's your best day, you know, and the first week you do it properly makes your last week so much easier – no lame cattle, nothing, you know – no trucking and all that.'[81] Arthur McRae was one of the few drovers still working into the twenty-first century, and his philosophy of starting the drove well in order to finish well held him in good stead for years. Those drovers who had the skills got the lion's share of the work, while the 'hoolers' and 'cowboys' did not tend to last the distance.

Sometimes circumstances dictated that a drover might need to slip away from a long drove to attend important events. Arthur McRae laughs at the memory.

> We used to do that when Hawke's Bay had the Ranfurly Shield – never missed a game . . . if you were going to the rugby . . . they wouldn't give you a paddock just to feed the cattle or anything, but if you had a good reason – like seeing the rugby – someone would give you a paddock.[82]

There was no manual when it came to droving; knowledge and skill came from observation and experience. Droving wisdom was accumulated by clocking up the hours, along with a word or two, in season, from an old-timer perhaps. No two days were ever the same on the road; but as the drovers perfected their craft, they had a better chance of avoiding disaster.

Sheep are transported over the Uawa River at Tolaga Bay, circa 1916. *Alexander Turnbull Library, 1/1-022067-G*

6. TRIALS A-PLENTY

Problems on the road

. . . when things are good it's a good life, but when things go bad you don't want to know about it.[1]

As we have already seen, a drover had to be ever watchful and prepared for any circumstance – usually involving conflict with traffic, difficulties with getting stock over bridges, trouble negotiating stock through towns, or perhaps the risk of stock poisoning. Such events added to the stress of the work, but at least when disaster struck, drovers were sure not to make the same mistake twice.

River crossings and 'the New Zealand death'[2]

> A Parliamentary return obtained by Sir E. Stafford showed so large a number of victims to our rivers that he stated drowning ought to be classed in New Zealand as a natural death.[3]

In the early years of settlement, when travelling with stock there was a high risk, for men and stock, of drowning at one of the many river crossings. Two such deaths are mentioned in Herries Beattie, *Early Runholding in Otago*: Robert Wilson, who drowned while fording sheep over the Mataura River on the Waimea Plains in 1859, and Thomas Newton, who drowned when crossing stock over the Taieri River near Sowburn.[4]

Drovers might spend hours, even days, at rivers trying to cross obstinate sheep; tenacity and a good dose of ingenuity were needed. Alexander McNab of Southland, travelling from Bluff to Invercargill in 1856 with bullock dray and sheep, 'put the dray into the creek with a pole on one bank, and formed a platform of timber on it so as to make a temporary narrow bridge, over which the sheep ran easily in their usual "follow the leader" style'.[5] In another instance a watertight wagon box made an impromptu punt for men, packs and horses.[6]

The approach and the timing of the crossing could mean success or failure. The wrong decision could be costly, as the manager of Lake Station, Charles Wilkinson, found out in 1908 when he tried to cross 3500 merino ewes at the Branch and Wairau. He arrived late in the day and made the

A mob of 400 cattle meander across either the bridge or the river, depending on their mood. They are at the Mokau River in Taranaki in 1924. *Sir George Grey Special Collections, Auckland Libraries, AWNS-19240320-40-5*

fateful decision to begin crossing the sheep near the junction of the rivers where there was a small island. Running out of light, the stockmen stopped for the night, leaving 500 ewes on the island. All were lost when heavy rain fell in the night and the river rose and washed them away. Further attempts to cross the remainder of the sheep were unsuccessful and they were left with little choice but to cart the sheep across by horse-drawn wagon, which took days. Some 800 sheep were lost.

Cattle, too, could be obstreperous when crossing rivers. Some stockmen used the 'ringing' technique to get them across.

> Stockmen found that the best method was to start the mob 'ringing' and when it was rotating like an enormous wheel, a man would suddenly ride into it, against the current of cattle, and using his stockwhip freely, break the ring. A few beasts would be shot outwards and swing towards the water's edge. The oncoming, rushing animals behind would push them still further, and the pressure continuing, they would be forced well into the river. Once there, they would look anxiously at the opposite bank, then, mindful of the barrage of cracking stockwhips behind them, plunge forward and swim across without further protest, the whole mob following after them. This work needed good men and good horses.[7]

Even once bridges were built, it was not always that simple. Cattle often had a deep suspicion of bridges, particularly wooden ones. The resourceful drover had strategies in place. Some swore by taking tired cattle over at the end of the day, believing they were less trouble then. Arthur McRae had his own trick at the Frasertown Bridge:

> . . . they put a bailey bridge over the Wairoa River – it's a big span – of course they'd clang and rattle, and an old Maori chap had a dairy cow there and . . . I used to let his dairy cow out and whack it on the bum, and over the bridge it would go and the cattle would follow it no trouble.[8]

If the cattle got spooked, the drover had to be ready and know just the right amount of pressure to put on them to avoid having stock go over the side.[9] West Coast farmer Pat Kennedy's solution was to:

> . . . get them running cos they were long narrow bridges and you didn't want them stopping on the way. You had to make sure they were going forward when they came to the bridge. Once they got on the bridge you were pretty much right. If they didn't get on the bridge or baulked or anything you were in real trouble.[10]

Fred Cowin, a Golden Bay farmer, developed his own approach to crossing fat lambs across a river:

> There was a spell, quite a spell . . . we drove fat lambs, we were told we couldn't drive fat lambs but we never had any real trouble. There was a bit of science about getting fat lambs to enter a river – you need three people, and two of them go forward and one stays back, and you get a small mob in front and you gallop them into it, and then you bring the rest of the mob up before they can get out. And we never drowned any doing that, and we did that for seven or eight years . . . There's craft in all these things you had to have learned.[11]

When things were not going quite so well, a drover might be heard for quite some distance reading the stock their pedigree. Nurses at the maternity home in Collingwood were often treated to such discourse when, on frosty mornings, Jack Richards was having trouble getting cattle beasts over the Aorere Bridge.[12]

Sometimes it was easier to drove cattle through a river than to take them across a bridge. They could get a good drink at the same time; and it showed consideration for motorists. Once on the bridge, though, the drover had right of way – and this often led to an altercation between irate motorists, frustrated with being held up, barging onto the bridge, and a just as irate drover insisting on their right to cross unhindered. Jack Curtis laughs at the memory of Ned Macky and him crossing the old single-lane Whakatane bridge with a mob of cattle and a big car barrelling onto the bridge, tooting its horn. Ned was seen rearing up on his horse, swearing and telling the driver:

> 'Get that car off the bridge! Get that car off the bridge!' And the girl, she might have been a teenager, I don't know, she – cos Ned swore a bit – 'Don't you let that Maori talk to you like that, Dad.' Well I seen Ned undoing his oilskin, well, that's the last thing, eh. 'Get that car off the bridge!' Well the cattle are all jammed up – well he starts walking the cattle on the back of the oilskin. Well they went over the bonnet they went through the bonnet what a mess. Jesus! And I was saying to Ned, 'She's right mate. She's right.' 'Bloody Pakeha bastard!' Well this joker was getting the police – he was doing all sorts. Never saw him again eh! And all the stock went across no hassles.[13]

Bim Furniss often had to cross the Huntly rail-and-road bridge. He remembers drawing the wrath of the local taxi driver.

> One day in particular the local taxi driver spent a lot of time in the pub and occasionally he'd go out and get a client. He had to come across the river but I was already on the bridge with a big mob. And he drove into them . . . he thought there was a . . . gap on either side of the train line . . . He got halfway into the mob . . . One of them ended up sitting on top of his bonnet. He never liked me after that.[14]

Sam Ruddell, a drover–landowner at Parakao, encountered an unusual problem crossing cattle across Mangakahia River in Northland.

> The Mangakahia River at Titoki was spanned by a swingbridge . . . Sam who knew the area well decided to take his large mob of steers over the bridge. Unknown to him overnight painters had arrived to commence painting the bridge. The cattle were quiet having been handled frequently so the lead went on the bridge. With the weight the bridge began to swing. The painters became frightened and decided to descend which startled the animals . . . Sam called 'Stay where you are,' but no, all they wanted was to get to safety. The result was the lead turned, jamming the way for the oncoming, which caused disaster. Some jumped over the rails, others blocking any further clearance were trampled on. The bridge swaying dangerously.[15]

There were no exceptions for VIPs, either. Don Monk remembers standing his ground when crossing stock over a Canterbury bridge:

> I was going to cross the Ashley bridge up here and two big flash cars come along and they pulled up behind me, you see, and they all had suits on and dark windows on one of the limousines. And he said, 'Hey, can't you hurry these cattle up?' I said, 'I know the rules of the road,' I said, 'when you're taking stock across you have right of way . . . and I'm not hurrying up for anybody.' He said, 'Do you know who we are?' I said, 'I don't care who you are, I've still got the right of way.' Apparently, it was old Sid Holland going home for the weekend. Old Sid must have said, 'oh leave him alone,' you know . . . All I was worried about is that he was going to run me dog over when he got up the road a bit.[16]

Tourists could cause trouble, too. Ray Stevens, from the Waikato, remembers the chaos caused by tourists trying to photograph stock coming over a one-lane bridge. He was busy trying to wave the tourists away when '. . . there was a clatter of hooves behind me and the girl that was in the team came racing up, and she said, "It doesn't matter whether they're from Norway, Japan or Canada, there are two words that every nationality knows: F— off!"'[17]

The drover needed to take care when crossing rivers. Some would slide off their horse and hold onto its tail as it was swimming across. Laurie McVicar, as a lad, was looked after by the older men: 'My neighbours, they were pretty careful, and if a river crossing was bad I'd get on behind someone else and lead my horse.'[18] Erle Riley was far from happy crossing the Parapara River in Golden Bay. Despite his fears he got to the other side safely:

> But on the way down the beach at Collingwood I'd go along there and I'd get a good start before many were awake in Collingwood. Of course there wasn't the traffic in those days, but there was the river, the Parapara River, you had to coincide with the tide . . . I remember one time something went wrong with the timing, somehow this jolly river was up, well, I got a bit agitated about that. I wondered what the hang I could do about that. Of course, to get them back to Collingwood wasn't an option. Anyway, I thought, I'll have to carry on. I'd read these stories of these drovers and that if your horse is swimming you slide off its back and grab hold of the horse's tail . . . I chased the cattle into the . . . I had the old stockwhip . . . they took a bit of getting into the water but eventually I got them in there, and then I had to go

in after them. I went in and I hung on like grim death wondering what was going to happen. And anyway the old horse started to paddle, it got too deep for him and I sat there as still as I could and he climbed out. Anyway I went happily on after that – went on down the beach.[19]

There was fun to be had at river crossing, too. Max Dowell gleefully describes his dunking when he was assisting on the South Westland drove. He was riding his horse and leading the packhorse when he got into bother:

My little horse Fairy had been caught in quicksand at the mouth of the Maori [Blue River] and when she got sand up under her belly she'd lay down, and when she got water up round her belly she'd lay down, quiver and lay down. I only had one spur so what was happening

Cattle are loaded onto a punt in North Auckland in 1902. Prior to bridges being built, stock were often transported by punt across rivers and waterways. The challenge for the drover was to prevent river-crossed stock from wandering off while bringing the remaining mob across. *Sir George Grey Special Collections, Auckland Libraries, AWNS-19020410-9-1*

there, if I was quick enough, I'd just give her a tickle with the spur you see and that was it – away she'd go.

Well for this crossing, the Maori, I was probably getting a bit tired on it or something like this, and I left it a little bit late and she stopped, so I tickled her a little and she took off – but the packhorse had stopped. Next thing between my bridle and the packhorse's lead rope I'm strung out, ripped out of the saddle, strung out there. My weight – both of them going out in opposite directions, my weight [was] too much for them, so they both come forward and when I went into the water with a splash they both reared back. And Dick Eggeling is up on the bridge, well, I thought Dick was going to have a heart attack, he's egging me on and egging me on and laughing and laughing his head off. Till I got the two horses side by side and then I was right, you see . . . Wet through, cold and wet through.[20]

For those droving sheep, bridges provided an easier way to negotiate rivers – although there was still the challenge of sharing the road with vehicles. This mob is on a bridge near Nelson in 1932. *Sir George Grey Special Collections, Auckland Libraries, AWNS-19320817-45-2*

Bruce Ferguson of Kaihoka droving sheep along Westhaven Inlet in the 1940s. Those droving stock near the coast often had to negotiate river mouths and tidal mudflats. It was imperative to time the tide. *Bruce Ferguson Collection*

Dealing with traffic

The advent of the motor vehicle brought another challenge for the drover. Initially there were few vehicles on the roads and their speed was limited – some went only 12 miles (19 kilometres) per hour. As vehicles became bigger and more powerful – and more numerous – there was more competition for the road, and increased conflict and vitriol between motorist and drover. Even when motoring was just getting under way the battle lines were quickly drawn.

PITY THE WORRIES OF THE DROVER

Motor cars have cast quite a terror on those who drive stock along our highways. One well-known drover informed a *STAR* representative that he intended relinquishing his present avocation as soon as a vacancy in another line of employment presented itself, as the responsibility now attached to the duties of drovers had become considerably increased of recent years owing to the ubiquitous motor car. At one time, a drover could follow a mob of sheep and read a paper or delve deeply into the pages of the latest novel; but owing to the reckless manner some chauffeurs approach livestock, a drover nowadays has to be continually on the alert to so direct his charges as to minimise the possibility of stock, sheep particularly, being run over . . . Our informant stated that one day, whilst driving a mob of sheep from Bulls to Sanson, no less than 27 motor cars traversed that four miles of roadway. We think the motorists should consider the troubles of the drover, and ease up when approaching cattle or sheep on the highways and byways. Consider the other fellow when he droves if you would be considered when you rove.[21]

TO THE EDITOR OF *THE STAR*

Sir,—You give us the complaint of the drover in reference to the motor car, but how about the motorists' side of the question. The road is blocked with sheep and the drover is strolling along, not taking the least trouble to as much as put his dog round them or clear a way through. I think the motorist has quite as much to complain of as the drover, particularly as the drover can take his sheep off the road and the motorist cannot take his car off. Presumably, the motorist has as much right to the road as the drover who blocks the road without the least compunction or attempt to meet other users of the road halfway. I doubt if there is any greater road-hog than the drover.—I am, etc., Motorist[22]

By and large drovers aimed to be considerate toward motorists. They had various techniques to avoid congestion. They drove along back-country roads where possible, to avoid the worst of the traffic; or they worked the stock along at just the right pace to allow vehicles through. It was not a matter of pushing the stock: 'if you start pushing the first thing that happens is all the tail closes up and you block the road and so then you've got a build-up of traffic that can't get through'.[23] By allowing them to drift a bit and 'mooch along and graze' they were often on the sides of the road and traffic could wend its way through with little problem.[24] Alex Cocks recalls the advice of an experienced drover: '"Make any cars on the road coming toward you or going, make them always pass on the one side," he said, "don't let them start going into your sheep from both sides. After one day on the road your sheep will get that way, they'll just drift when a car's coming."'[25] Alex is emphatic that, with repetition, the sheep got to know what to do.

Dogs, of course, were the drovers' greatest tool and could be called on to whip up the side of the mob to clear a path for traffic.[26]

Most drovers had a tale of a narrow escape with vehicles; and they spoke with vehemence of the stupidity of many motorists, and the lack of consideration for a mob on the road. Those who were out in the lead talk of trying to warn and slow motorists down by frantically waving, only to have the driver of the vehicle mistake the wave for a friendly greeting and cheerily wave back as they careered into the mob. Gill Pullen remembers being out in the lead trying to slow down a motorist on the main road into Huntly and watching as her husband – with horse – was skittled.

> It was just horrific. I mean, one car there – I was up in the lead – I tried slowing it down; it was just no way. And it just went barrelling at one speed right through the middle of the mob and out the other end. Well Bill could see him coming and he was trying to get the dogs off the road and thought, 'Well this fella's going to do some damage here,' and he sort of pushed his horse into the mob to open the mob up and this fella [went] straight in and underneath his horse. Here goes Bill, his horse and everything . . . I could see it from the other end and I thought 'MY GOD!'– and I couldn't leave the lead to go back but anyway – next thing I see him back on his horse and I see this car take off. I mean they didn't even stop.[27]

Old Dollar Murray on Ken Lewis's Northland drove had perhaps the most original method of arresting speeding traffic. Phyllis Clements still laughs at the memory of his antics.

> We were coming up the Herekino Gorge and Dollar was with us. Anyway this car came flying down the hill and Dollar he started to do the haka – and the tongue – and he was doing the haka and he'd sort of get down on the road and crawl toward the car – I'll never forget it – and these poor people nearly died of a heart attack – ground to a halt, I can tell you. Oh, it was funny.[28]

Another time for Gill and Bill, tragedy was averted just by sheer luck. Going into the Matahina one time they encountered a milk tanker coming 'off the hill over the dam when, hello, here's a whole heap of cattle':

> . . . and he jumps on the brakes and he's got no brakes. And he ended up going on top of all these cattle. Well it comes out onto a . . . flat . . . and I'd just got them out and was approaching this gully going up this road and had them all sort of basically blocking the road getting them to draw up, and down comes this truck and he comes straight into them. And I thought, 'God,' and he comes straight into the front. And the front was reasonably open, but the tail of the mob, of course, because

Negotiating traffic could be a headache for the drover. Here, a drover on a highway in the Wairarapa in 1945 works to control his mob. Note the replacement of a horse with a pushbike. *Alexander Turnbull Library, 1/4-001293-F*

I'd been putting a bit of pressure on to get them to draw up this road which closed up, so he run over them. He virtually run up and it lifted the truck right off the ground; it was an articulated truck, so three axles. Well he rode up over these cattle and he had three or four and I can't remember – underneath the truck. He'd finally come to a standstill. And the poor chap in the truck was terrified, trembling actually. And I rode across rather rapidly – and he got out of the truck and wanted to know 'What do I do? What do I do?' And I said, 'Get in that bloody truck and back it off 'em.' . . . He leaps in the truck and backs back off and these three or four cattle beasts get on their feet, shake their bloody heads and tails and carry on towards Mercer. One lost his tail – a fraction of his tail was torn off under the wheel of the truck. But they just got up and walked away. The poor fella, he was in a hell of a state.[29]

Even when motorists were doing everything right, events beyond their control could lead to near-disaster as North Island drover Noel Martin could attest to.

> I can remember the roadman coming along on his pushbike. And the cattle are not used to a pushbike and he had the shovel strapped along the bar in the centre in the old roadman style, and there was this old couple and they were very quietly and very nicely doing the job all proper and they just started going into the stock and this chap came in the other way on his pushbike and there must have been one beast hidden behind the other and didn't see the pushbike coming and all of a sudden he saw it – well, he went clean across the bonnet of this car sideways. And this poor old couple, they were shell-shocked. There was one in the back: 'What happened? I didn't do anything,' and I said, 'No you didn't, you were just really unfortunate.'[30]

Drovers were no saints, though, and when push came to shove, they were not above defending their right to the road. Those impatient motorists who tooted or pushed impatiently through a mob, the drover would give them 'a bit of a rattle up'.[31] Their form of retribution was to draw the car into the midst of the mob and then innocently look as if they were guiding the dogs to assist the motorist, while all the time quietly preventing the motorists from extricating themselves from the mob.

> I can remember my uncle cursing these idiot drivers that would drive through a mob of cattle and scatter them all over the place. But then, the drovers would have their own way of handling them too. They'd sometimes let a car go in the same way as the cattle . . . they'd get the leading dog to hold the cattle till the car's halfway through and then they'd shift the leading dog on and the cattle would go on and the car would be in the middle of the mob and they'd bark their dogs behind and make out that they're trying to help. And they'd whistle the leading dog to bunch 'em up. And some of those cars come out of it, although they were heavier panelled cars than they are [today], some of them come out with quite a few bumps. But it only looked like an accident. And one of the favourite places was on the Reefton saddle – south of Reefton – because the road was a lot of horseshoe bends. Although it might only be 100 yards from where the drover was to the lead of the cattle, it was only 20 yards across the gully so that he could whistle his dogs up and put pressure on. And sometimes they'd walk a car along for miles. You know I heard of one case – the car scattered cattle everywhere in the morning, and they were still on the road with

the cattle in the afternoon, so this guy arrived back – so they put him
in the middle of the mob and took him along. And some of those cattle
would just about climb over the mudguard of a car.[32]

Such sport would have brought some satisfaction to a drover who had
nearly lost a dog, or who had had to listen to a tirade of abuse. And of
course, vehicles that underwent such treatment in a mob of cattle would
later have a panelbeating bill. When compensation was sought, Jack
Richards had the ultimate response: 'Ask the bastard that did it.'[33]

Trouble in town

Inevitably droving through town – no matter its size – had its share of
problems, including barking dogs, children playing, a woman's dress blowing
in the breeze, open gates, a tempting shop, or just stroppy animals. Younger
drovers watched and learned from older drovers who had run the route
before and knew the pitfalls. Planning was essential: arrangements were
made where necessary for extra hands to come and help get the stock
through town. There were always extras keen to help – young lads who
wanted to work with stock, or other local drovers. Stock routes skirted
around towns; but routes to the saleyards, abattoirs and freezing works
often directed stock right through the main street. It was not unusual to see
cattle or sheep slowly meandering past the Fitzroy shops in New Plymouth
on their way to the freezing works at Waitara in the 1940s; and though
it's hard to imagine now, stock negotiated Greers Road, Memorial Avenue,
Ilam Road, crossing Riccarton Road into Middleton Road and headed down
Blenheim Road into Christchurch's Addington saleyards every week, right
into the early 1950s.[34]

Today it's hard to imagine prize petunias being grazed by hungry stock
as they moseyed on by, let alone residential streets covered in excrement,
but it was all par for the course in a young country, with strong links still
to the rural life. Not that it didn't cause conflict, even back then. People
who lived on the stock routes knew to keep gates closed – although not all
did; and while drovers tried where they could to prevent stock going into
properties, inevitably the odd one got away on them. Riding in after a beast
– ducking trellis, clothesline and any other backyard hazards – the drover
worked frantically to reclaim the errant animal, hoping desperately that the
homeowner was not present. You never knew how events would unfold:

> There was a cow that walked along the footpath . . . this old cow had
> seen a gate partially open and pushed her way through the gate and

down the pathway to the side of the house where she was confronted with the outhouse. Well the door of the outhouse must have been slightly ajar and the old cow must have thought that was the door into the stall so she pushed on and went through – the floor gave way and she crashed through. At that time the householder was disturbed and come out. And we have a description that a newlywed, the householder was a newlywed, her husband was working at the time and she come out and she was dressed in what was . . . a see-through arrangement. That was the least of their problems: this particular lady was well versed in a line of language that the drover hadn't quite caught up with at the time, and was able to express this very, very colourfully. And [she was] exclaiming: What was she to do when the old man comes home at night, that was the first place he would go to – what was she to do? . . . But the drover, my father and his assistant . . . eventually got her [the cow] out . . . with the help of the neighbour.[35]

Drovers faced all sorts of issues when travelling through towns, one of which was the open gate. Those on a stock route quickly learned to keep gates shut when stock were being driven past. This large mob of sheep from Five Forks and Fuchsia Creek are being driven down Severn Street in Oamaru in 1923. *North Otago Museum*

Efforts to circumvent any damage to property were not always appreciated by homeowners, and drovers were left to face their wrath. Poor old unsuspecting Northland drover Billy Povey was the recipient of some rough treatment by one irate woman.

> Billy Povey was with us, he was only a little fella. As you went out of Kawakawa this woman had taken over the roadside – it was a fairly steep bank and she'd planted it all in flowers – at the right time of year it would look quite pretty. A cow had gone up there and Billy was nearly going to send his dog up and then he thought, no, it would just make the beast come skidding down the bank and make a hell of a mess. So he went up on foot and he's working it round nice and quietly and the next minute there's a WHACK across his shoulders and he looked round and here was a woman there with a broom in hand and she was ripping strips off him and hoeing into him with the broom. So he thought, 'blow this', and he put the stick on to the beast and the whole lot of them just skidded down the bank. But that woman was not happy, she was not happy.[36]

Other times a family might come home and truly not know what could have happened to their property.

The drover also faced the challenge of negotiating urban traffic. In the 1930s there was still a close urban–rural link, as shown by this image of sheep being driven down Glenmore Street near the Botanical Gardens in Wellington. *Alexander Turnbull Library, PAColl-3071-1-07*

Old Jack he liked me to go, he'd ring up, you know, and say, 'Well, could you come and help me down the hill?' because there were lots of places that the cattle could get off. Sometimes I'd go ahead and if there was an open gateway . . . I would stand there and then perhaps go ahead again and things. But one day we were taking the cattle up to the paddock in Lower Moutere – we went up Parker Street in Motueka and then we went along and up to the top end of Poole Street and there was someone there that was building a pipe fence and they'd got all the uprights all along on this fence – there was no gate or anything – they obviously hadn't finished it and they had the top rail along on top of these uprights. Well one of these great big cattle went in the open gate, walked along and then decided he was being left behind and got out underneath this pipe and the fence went along a little way and then it had a big bow in it. I don't know what the people would think when they came home and saw that.[37]

There were times, however, when the unsuspecting public were the recipients of some rough treatment themselves from an irate beast. Local newspapers are the repository of many a tale of man (or woman) versus beast, and in some cases legal action was taken against a drover.[38]

AN IRATE BULL
LADY ATTACKED IN THE STREET
Christchurch, Oct. 31

Attacked by a bull which displayed more than usual ferocity, a young woman had a few anxious moments in Colombo street, last evening, and her equanimity was not restored until the animal was beaten off by a party of men who quickly realised her predicament. She was unhurt except for a few abrasions and the shock.

The bull, which was one of a herd in charge of two drovers, was being driven along Colombo street near Dean street, and its restlessness was causing the men in charge not a little anxiety. Eventually, after several futile attempts, the animal was successful in separating himself from the remainder of the herd and set off in a meteoric, if unceremonious, retreat along the thoroughfare.

The young woman had just stepped out of a neighbouring store when the animal sighted her. The bull came to an abrupt standstill, and stood gazing at his victim; then, with a hysterical cry and a

> fluttering skirt, the lady made a determined sprint for a nearby paddock. The young woman covered the first 25 yards in remarkably fast time, but the bull had four legs, whereas she had only two, and the pursuing animal had overtaken his victim before the 50 yards' mark was reached. He shot past her like a flash, pushing her hard against the fence . . . things were . . . dangerously near a climax when several men with sticks rushed in, and . . . the animal was prevailed upon to resume his place among his fellows . . .[39]

Murray Dymock remembers one fool standing his ground when stock got away at the Addington saleyards.

> We used to have some strife. There was a whole bunch of them out playing cricket one night and we'd bought some Chatham Island cattle – course big five- to six-year-olds with horns about four foot wide. And the whole mob went through the fence out into Hagley Park. And course these guys [were] out playing cricket. One of these must have been off a farm, he was going to stand his ground. And I'm yelling at 'em, 'Get the hell out of it.' These Chatham Island cattle hadn't seen a man on foot. And here's me trying to get to the lead and turn these cattle on this grey mare I had. And this guy, 'Oh no, I'm alright.' And I said, 'Get out of the bloody road.'[40]

The drover deserves a certain degree of sympathy in such a case. However, while it caused a moment of intense stress for Murray, it all ended without mishap.

Northland farmer Allan Crawford laughs as he tells the story of one wet day droving past the Te Kao meeting house.

> There was a funeral on – they had a flag flying at half-mast and the wind was whipping it and cracking it and it was fluttering, and then the other side of the meeting house the children were playing and their screams and yells were coming through on the wind to reinforce this cracking and whipping. And the cattle didn't want to go past. So after a bit of hard work, I got the front end past. I took about 35 past. And when I was past it I relaxed and turned around and here's the rest of my almost 700 cattle turned and going away from me . . . The last man, Kenny Whittaker – Skip's son – had good dogs and held them and till we regrouped and got them past.[41]

The main streets were also the scene of exciting events – whether it was sheep or cattle perusing the wares of the local haberdashery, or perhaps a stray animal diverting to the local hostelry to quench an insatiable thirst, you could never be sure what might eventuate. For one poor beast, the lure of its reflection in a barbershop window enticed it right through the glass.[42]

> The life of a sheep drover is not always the serene one that some would imagine, and those persons in the Avenue about 4 o'clock yesterday afternoon who witnessed a mob of sheep going through, and who were previously unaware of the trials of droving must certainly have realised this. The mob, totalling close upon 1000 woolly backs, soon got into confusion among the trams, motors and other traffic in the street, and gave the drovers and their dogs a lively time till they reached the bridge. Before reaching it, however, a few lively incidents happened, sundry members of the flock breaking away and darting into the shops. One of them found its way into Messrs Buckrell and Co's draper's shop, and after a bewildered inspection of the goods, was eventually convinced that its fleece was not just then required. When ejected it was alone, and persuaded to rejoin its mates across the river . . .[43]

While it's difficult to see any silver lining in such encounters, one rouseabout in Greymouth was only too keen to see the unexpected visitation of errant bovines in the local hotel. Owen Gibbens's father was driving some stock to the local abattoir in Greymouth. As they passed the Gilmore Hotel, one old cow diverted into the establishment.

> . . . so she went down through the hotel and out through the back door. On the way she left a visiting card and there was some discussion with the proprietor and the rouseabout, and the drover managed to get the cow back out and they proceeded with no further trouble. However, the next time the drover, my father Dick, come past, the rouseabout made a call: 'Hey Dick, when are you bringing more cows through?' . . . He said, 'It was very good, I got 10 bob for cleaning it up.'[44]

There were times when the townspeople took exception to stock coming through their town. Keith Ribbon had to do some quick thinking when he was confronted by some of the locals at the Tolaga Bay bridge:

> Some of these people from town had had enough of cattle coming through the town and they were going to stop it. So they got on the

bridge to stop the cattle. So Keith rode up there, he was on his own. He said, 'Are you going to get out of the way or not?' They said, 'No.' – 'Right!' So he went back and got about 30 cattle and just went in, and . . . he pushed these cattle, he got cattle across the bridge like this and pushed these people out in front of them. So then he turned round to go back and pick up the tail . . . In the meantime, the leading dog swum the river and shot out round the lead and these people came onto the bridge. And as the tail was coming onto the bridge, the rest of the cattle, the lead were coming back and it started a concertina effect. And Keith said, 'Have you ever seen 15 people go over the side of a bridge into the river?' He said, 'That was the first and last protest on the bridge in Tolaga Bay.'[45]

Further south, Noel Martin also encountered opposition, this time from tradesmen protecting their work.

Coming through Wairoa you'd come over the main city bridge – they blocked that in the end. I can remember Frank Lambert giving us a hand – he lived in Wairoa, and Linc and myself taking this mob through. They went through as a group, they weren't strung out at all, but two weeks before or a week before and they had the boxing for the new motor garage on the south side of Wairoa on the left, they'd had the boxing up and they'd brought a big mob of cows through and they'd smashed the boxing and everything. And we went through . . . it was dark when we let them out of the paddock and we were trying to beat everybody . . . It was barely breaking light as you hit the bridge over the main Wairoa, and standing all out – word was out apparently – standing right along was carpenters and tradesmen waving leather aprons – they weren't going to have their boxing destroyed again – which you can't really blame them.[46]

Noel and Linc's efforts to calm the situation had little effect, and they had to deal with animals spooked by the uproar.

One tactic for getting stock through town was told to Laurie McVicar by his Uncle Bill, who took stock over the Otira, before the tunnel was built, and loaded them on rail trucks at Arthur's Pass. In order to get the cattle through the townships they would give the cattle a day off and have the mob in a paddock. Borrowing some extra horses, the drover would spend the day training the cattle to follow the horse. When it came to going through town they would 'just follow something, they'd follow a dog or anything that moved, it just gave them confidence'.[47] Like any plan, though, there was always the potential for it to go awry. Laurie's uncle talked of the day when this one mob obstinately refused to move until the old grey horse they had followed the day before came out and led them through town.[48]

A drover would say, without hesitation, that it was the stupidity of people that made their job more difficult. It was here that the strongest element of the drover's arsenal would come to the fore – their voice: when all else failed, a bellow from the drover was often all that was needed to clear the path. The wrath of some drovers was legendary. John Sullivan from Fox Glacier stills laughs at the story of Billy Condon's skill at enabling his stock to move through town unhindered.

> Harihari was where the north bus met the south bus and they had lunch and drivers changed over . . . and when we got up to the township the buses were just about ready to depart. There were people who saw the cattle coming and they were standing around and looking . . . girls working at the hotel had their heads out the windows. And the Road Service garage was just across the road from the hotel and there was a clerk there, I think he was fairly new to the district. He came marching up the road there with his camera to take a photo. And, ah, Billy Condon was quite a character. Billy's first utterance was: 'By Christ, some people rear stupid bloody children.' He expanded a little bit more and it drew a bit of doubt on the parentage of this guy, and he then told him where to go and how to go, too . . . The cattle were all milled up in a mob . . . This guy took off as hard as he could lick it, the tourists started streaming into the buses and the girls pulled their heads in the windows and the cattle just walked quietly on through. It was brief and blue.[49]

Tutu poisoning

Drovers have always had to watch out for tutu, or 'toot' (*Coriaria arborea*) – a plant that is at times poisonous to stock. Francis Hamilton (later Mayor of Greymouth 1877–78) was shocked by the impact of the effects if tutu on his just-landed sheep.

> Last Saturday we landed the sheep and all over the Country here is a poisonous shrub called Tute which affects all sheep newly landed. Sunday nearly all of ours were tuted, they are taken with spasms and lockjaw and seem to be in awful pain. We Bled nearly all of them and gave them ammonia, but in spite of all we could do we lost thirty of them . . .[50]

Samuel Butler, the successful Canterbury pastoralist and author, offered his insights into tutu in his account of his first year in New Zealand.

> Tutu is a plant which dies away in the winter, and shoots up anew from the old roots in spring, growing from six inches to two or three feet in height, sometimes even to five or six . . . The peculiar property of the plant is, that though highly nutritious both for sheep and cattle when eaten upon a tolerably full stomach, it is very fatal upon an empty one. Sheep and cattle eat it to any extent, and with perfect safety, when running loose on their pasture, because they are then always pretty full; but take the same sheep and yard them for some few hours, or drive them so that they cannot feed, then turn them into tutu, and the result is, that they are immediately attacked with apoplectic symptoms, and die unless promptly bled. Nor does bleeding by any means always save them.[51]

Other points not mentioned by Butler, but pointed out by old drovers, were that frosted or wilted plants and new growth seemed to be more toxic to the animals: 'The frost affects it . . . and you only need a couple of leaves . . . They swell up to hell. It's the toxicity of it that kills them.'[52]

Cattle going through the Waioeka Gorge apparently half lived on tutu in winter time:

> . . . certain times of the year that tutu is dangerous or if it's been cut, if someone's cut it and it's wilting, then it's lethal then . . . otherwise if you're going through your cattle are just reaching up and chewing away at tutu all the time – you learned not to panic about it . . . you accepted it but you were always alert . . .[53]

Bleeding was the common remedy, despite some scepticism as to its merits: 'They said bleed them, bleed their ear, or cut the roof of their mouths, aahh, but I think that was an old fairy tale – I don't think it really cured them.'[54] Opinions varied, too, on where to bleed, and on why it worked best:

> Then you get busy with a knife, you either cut the tip of the ear or tail anywhere to make them bleed. Run a jagged stick up their nostril and that will relieve the pressure on the brain. And sometimes it will save 'em.[55]

> . . . nicked them under the tail or behind the ear and bled them for a while and they'd be alright.[56]

> . . . through the vein that runs from his nose to his eye . . . with my knife . . . A lot of people [think] . . . the beast is actually drunk, he . . . gets the staggers and they rush up and bleed him – they just cut him, they cut the vein under his tail. He doesn't get any blood pressure in his bum, he gets it in his brain. And my old man, I've seen the old man do, oh, 30-odd one night and he never lost a beast.[57]
>
> Up the nose is the best – broken stick, dry stick and break it – make it jagged and get their head – it sounds cruel – but just rub it up their nose – that's where you bleed mostly from – the blood just, whoosh, you know . . . What they die of is it builds up pressure in the heart and if you can release that pressure that is the best way . . . they just blow it out and there's no sign of blood – cut their tail then there's blood around their legs. I've never cut their ears . . . their tails are second best.[58]

Pragmatic as always, drovers got on with the job at hand whether they liked it or not. Don Monk remembers one trip where a number of his mob of cattle went down with tutu poisoning:

> Oh yes there were some trips, like, I bought a mob . . . we came into a big tutu patch and the cattle were getting tired. And there were about 20 went down with tutu poisoning . . . they go mad like they're on drugs or something. So – a joker told me either cut the roof of their mouths with your pocket knife or cut their tails off. So – I bobbed them all, just below the meaty part of the tail. Of course, all these cows are getting along the road with bobbed tails and everyone wondering . . . I only lost – I saved them all except two. One jumped over a cliff and drowned herself in the Waiau and the other one just kegged on the side of the road, so the grader driver came along and he pulled him off the road for me. That was one of the days I thought, 'Oh I'll never go droving again.' It was real upsetting.[59]

Mike Holland, as a teenager droving with Ken Lewis in Northland, laughed at the memory of a 'tutuing' incident where all of them were off their horses bleeding cattle using their pocket knives when along came Ken with his butcher's knife. Ken, an amputee with a wooden leg, got off his horse and wasn't quite balanced right, 'and he was struggling on his good leg and he lifted the tail up and he had this butcher's knife and he went schhh – and the whole tail came off.'[60] Mike joked that there wasn't even any oxtail soup for tea that night.

Drovers had different strategies for avoiding any tutu incidents, such as punching stock hard through known tutu areas so as not to allow them time to feed, or letting them have time to fill their stomachs before passing through bad areas. They had to be alert to the problem – Alex Cocks always 'put a dog around the side of it' and hurried them past, not giving them time to feed.

While spectators to a drove may well think the drover was doing little more than following behind a mob, the drover knew he was earning his keep. As the saying goes, 'An ounce of prevention is worth a pound of cure', and for the drover that meant their eyes were constantly peeled for any sign of trouble.

A drover with his horse and gig on the road near Kaikoura in 1910. *Sir George Grey Special Collections, Auckland Libraries, AWNS-19101013-10-2*

7. PARTNERS ON THE JOURNEY

The drover's horses and dogs

. . . having a flash horse and the team of dogs, and all the young fellas in Kaiapoi thought I was great riding me horse through the town with me dogs following me, you know what I mean . . .[1]

For the drover, the horse and dog were like hammer and nails to the builder or chalk and blackboard for a teacher. They were essential resources and without them the drover would have been lost. Each drover had preferences for what they looked for in their animals, which made for quite a kaleidoscope of mutts and nags. No matter their look, however, they were on the road to do an essential task, and if not up to the job there were few chances of redemption.

Horses

'You better get back on or I'll kick your arse' – that's how young Barry Grooby's riding lessons began.[2] Learning to ride was a rite of passage for country children in New Zealand. The love affair with horse and dog often started young for those on the road: girls and boys were already riding ponies or some 'broken-down old nag'[3] at the age of four or five – more often than not bareback or just with a sheepskin, clutching tightly to the mane and hanging on with the knees.

John Holland, a Northland drover, recalls overhearing his father in conversation with his mother one day: '. . . he said, if the doorway was big enough he'd have his horse at the table for breakfast.'[4] And Laurie McVicar remembers pestering anyone who would listen to take him riding.

> I was always mad on horses, and I can remember getting along in our hotel kitchen with a bridle waiting for someone to catch a horse so I could have a ride – and I wasn't tall enough to hold the bridle off the floor: the bit was dragging on the floor when I was holding the top end of it.[5]

Laurie recalls being taught to 'pull the right rein to go to the right and pull the left rein to go to the left', but as to any finer points there weren't any:

> It was more experience than tuition – you just did the miles, the
> hours. You found a place on a horse, by modern standards it might be
> wrong, but . . . you sat there and somebody said 'if you put your feet
> a bit further forward you'll be a lot safer' or do this or do that and you
> just did the miles and found a natural position to sit in.[6]

West Coaster Hugh Monahan recounts his unusual experience of learning to ride:

> Well I used to ride cows for a start. Our aunty always used to run cows
> out on the roadside, and one would have to go and get them. And I
> thought 'to hell with this' walking behind them. They were out on the
> roadside and they might be a mile, they might be two miles away. I'd
> mount one of them . . . It was better than walking.[7]

Of course 'Shanks's pony' was always an option for a drover; but the horse was an essential element in most droving teams. 'Hack and pack'[8] – the drover's steed and the packhorse – were a common sight alongside the moving mob. The packhorse, if it was well trained, was out in front, making the most of the opportunity to feed as it meandered along the stock route, and the hack was generally in the rear with the drover.

In early pioneering days horses were scarce, and Shanks's pony was truly the only option for those moving stock. As land was settled and stations became established, the demand was mainly for mixed-breed horses, good for saddle and light draught work. Invariably, they were a mix of thoroughbred, Clydesdale or draughthorse, with sometimes a bit of Arab and possibly pony thrown in. They were a strong, small- to medium-sized utilitarian horse, largely bred and imported from New South Wales.[9] Initially, perhaps, only the runholders would have had a horse, but by the mid-1860s horse numbers were increasing and 'the men often owned a horse of sorts, and when they left a job would depart on horseback with their swags strapped to their saddles'.[10]

Drovers preferred horses that were a 'good walker' or possibly an 'ambler . . . it's a fast sort of a walk – both legs from one side going – just like in a rocking chair'.[11] On the long trek home after a drove, with the dogs following along behind, the horse needed to go at a reasonable pace. Jack Curtis preferred a 'walker' over an 'ambler': the ambler's pace, at some 7–8 miles (11–13 kilometres) an hour, he felt was too fast compared to a walker's at around 6 miles an hour; the slower pace meant his dogs could keep up with them.[12] Of course, if the route home was on a rail route they could always hire a rail wagon and load horses, dogs, gear and self and cut travelling time significantly, while still pocketing the wages for travel days.[13] Don Monk, droving in Canterbury, looked for a good cutting horse – if stray stock ended up in his mob, he wanted a horse that could go into the stock and cut out the errant beast.[14]

Others were less fussy: Murray Dymock remembers the first horse he bought after leaving school:

> I bought an old trotting horse – I paid five pounds for him. He taught me to ride too – he'd drive flat-stick – a mouth like pulling on a gatepost. You couldn't stop him, he just went. I managed to stay with him . . .[15]

Later, when he was droving for Canterbury dealer Stan Wright, Murray had to regularly change horses, as it was possible to 'burn a horse out. It was a good horse that lasted six weeks. And he was really civil by the end of that six weeks.'[16] Stan kept plenty of horses at his workers' disposal, on land where Pegasus township is today. Murray would head down, saddle one up and 'take it for a gallop, take it down the beach or somewhere'. If he did not like that one, he would try another until he was satisfied.[17]

Young Rob Murray, keen to get out on the Ken Lewis Far North drove, had little choice but to ride whatever 'mongrel' Ken gave him; and just when he'd straightened the horse out, Ken would exchange it for another mongrel. On one occasion, Rob was given 'a beauty' and was not quite so ready to give him up. When Ken began to make noises about taking him away, Rob gave Trump – as he called the horse (he'd always wanted a Triumph motorbike) – 'a strop around and bumph off, he bucked me again'. When Ken saw that, he decided the horse was still a mongrel and Rob was able to keep his favourite horse a while longer.[18]

For Laurie McVicar, the move from riding his 12-hand pony to using his father's 16-hand ex-racehorse 'was like getting off a tractor into a Mercedes motor car'.[19]

Danger was ever present for horse and rider when dealing with horned cattle.

> Sometimes we used to take the cargo to the Thames, put the vessel on the beach, wait until nearly low water, then start discharging onto the beach, anytime in 24 hours. One night, we started discharging about 8 p.m. A good many were very wild. There were two stockmen receiving them on the beach. When about half of them were discharged, they broke away with the stockmen after them. It was somewhere about two hours before the stockmen got them back to the

> ship again. The two wildest were left in the hold of the ship until the last. Eventually [one] was got into the slings and hoisted over the side. Soon as his hoofs touched the shingle beach, he made sparks fly and straight through the mob for the drover's (Tucker's) horse and planted his two great horns in the horse's flank. It was that much injured that it had to be shot at once. Had it been daylight, it would not have happened, as it was a splendid little stockhorse and would have given the beast her two hard hoofs on his brain box and knocked him over, but being dark, neither rider nor horse saw the beast coming in time to get speed on.[20]

The horses had their quirks, of course: they might suddenly decide to take a roll in the sand while droving stock along the beach;[21] or steal bread from the farmgate mailbox; or try to run you down to avoid being caught. Drovers used some ingenious methods of thwarting them that, in this day and age, might not be so easily condoned.

> Ken had a grey gelding, it was hard to catch – it was a real mongrel to catch. And it would run over you even if you stood in front of it . . . And I'll never forget this morning . . . we tried to catch this horse and it just bowled us – down we went. And old Sid, he went and got a hunk of four-by-two. I said, 'What are you going to do?' and he said, 'You just chase the horse up,' and he waited and the horse came screaming up the fenceline and he waited, boom – chop – out. I said, 'Christ man, you've killed it.' 'No no,' he says, 'it's just concussed.' Well, when that horse got up he never ever tried to run past again. That was the end of that. And that's true![22]

Don Beard recalls his neighbour Charlie Sundgren's horse, Captain.

> Oh his stupid horse. In those days . . . you used to get hot bread delivered and we had this breadbox or mailbox . . . at the top of the drive. It had a good latch on it, but the damn horse seemed to know how to open the latch and had a taste for our bread. Good old Charlie would be screaming up the road on his pushbike – and you can just imagine him, I guess, his legs splayed and his arms out wide and his hat pulled down over his ears . . . He used to get embarrassed – we'd end up with just the crust.[23]

There were stories told of horses finding their way home without rider assistance.

A hand on an Oreti station, lost in a dense fog, took off his horse's saddle to act as a pillow, but acting on sudden thought resaddled and completely trusting to the horse's superior instinct it landed him safe at home although late at night.[24]

Davey Gunn raised cattle in the Hollyford then drove them to Lorneville saleyards. His horses could safely get him home, even in the dark and the extreme terrain. As Arnold Grey remembers it: 'Davey would simply lie along the horse's back, not even touching the bridle. When the horse stopped he knew he had arrived at the next hut.'[25]

Joe McLaren, droving stock through the Lewis Pass to Addington saleyards, could rely on his horse making his way back home through the pass while he and his dogs travelled home by Newmans bus – the horse waited patiently at each of the cattlestops along the route until someone let him through.[26]

Along with the 'hack' came the 'pack'. While the packhorse was not necessary on local droves, it was essential on longer droves, before the days

A young Harry Frank poses with his horses. *Harry Frank Collection*

The value of a good stock horse can never be underestimated when out on the road with cattle. This group of drovers and their mounts are on the Hurley drove in the Manawatu. Keith Sowerby is pictured second from left. *Julia Sowerby Collection*

of trucks and caravans. The ideal packhorse was a quiet, sturdy horse – not too big, though, as that made it difficult to lift gear on and off. A quiet temperament was key, so you could pack it and travel without incident. To have your packhorse 'setting the table'[27] – distributing the contents of the tucker box all over the road as it bucked its load off – was something all drovers wished to avoid. Harry Frank, droving in the central North Island, remembered one horse he was quick to get rid of: 'He was a mongrel – several times he dumped a load . . . You had to watch him all the time.' That same horse had actually put his previous owner in hospital. In the end Harry tired of him and sold him at the horse sale in Te Puke, not before he had advised one man of its unsuitability: 'A joker come up to me and asked would the horse be suitable for his wife. I said, "Do you love your wife?" He said, "Yes I do." I said, "Well don't buy it."'[28]

Arthur McRae liked to have two horses of similar height so he could interchange their gear: covers, saddle and packsaddle. His trick was to have both trained as a packhorse; he then alternated riding on one day and packing the horse the next. His reasoning was that the packhorse, once trained, was out the front getting plenty of feed, while the pickings were a bit thinner for the one being ridden; by alternating, he ensured both were well fed.²⁹ There were other tricks drovers used to ensure their horses were well fed: some unsuspecting cockie's paddock might play host to a drover's horse for the night, with the horse tied to a long line and hidden behind a hedge.³⁰

Packing the horse was an art quickly learned, often the hard way: if gear was not distributed evenly, the horse was likely to develop back problems; or the pack would slip under its belly and the horse would kick the gear to pieces. Bill Pullen had packing down to a fine art; years after his droving days had passed he could still recite his routine.

> Your pack slings just went around it and you'd do 'em up and then you'd put 'em on the hooks on the packsaddle on the packhorse – put your swag on the other side which was – being a chaff sack your sleeping bag, all your clothes, your suit, in the chaff sack and you rammed everything in there and you've got dog collar going round there, and you do the dog collar up so nothing falls out, and then you pack things around it. You put that on the other side. You put that on first – that's the first load to go on cos it's a lighter load. Your tucker box could be a bit heavy so that goes on the second load – . . . Then you put a top load on, what they call a top load which will balance up your load . . . when you put your top load on, like your dog tucker and your camp oven and stuff all goes on top just to balance things up, and you'll have what you call a pack cover which is six foot by six foot and you tie it all round . . . just a square of canvas six by six that keeps all your stuff dry . . .³¹

Even once the packhorse was superseded by the truck, some drovers were not so ready to give them up: 'Well the thing is, if your truck broke down or you had to go cross-country or road block or anything else you can throw your bedroll on your packhorse and swing a billy on the hook and you're right, that's all you need.'³² Arthur McRae kept his packhorse with him for a good couple of years after getting his truck, so he always had his back-up: 'Had the packhorse so you slept at night.'³³

A commonly held belief among drovers on the West Coast was that you never rode an east coast horse on the West Coast, as they were unused to West Coast rivers. Like any rule there are always the exceptions, however. According to Laurie McVicar his uncle bought two horses from St James Station in North Canterbury and they were fine as they had grown up in the Ada River catchment.

> And those horses were – oh – Old Stumpy was like a boat you know. I remember going into the Grey and the Robinson, between 'em, behind a mob of cattle once and my uncle was on Stumpy – he was a

The gig was a convenient method of droving: the drover could rest his weary legs, carry extra gear or a knocked-up sheep. Even dogs could be rested – in prime seats if they were quick enough, or on the sacking sling under the gig.

Occasionally a drover would also be seen out droving stock on a bicycle. Alec Sutherland of Murchison used a bike when he was assisting Charlie O'Brien, a fat-lamb buyer, droving lambs up to the railhead at Glenhope. 'I just had a bike.'[34]

When he got his car, Manawatu drover Keith Sowerby was able to squeeze in more work: 'Whereas before he could do one day's work in one, he could now do two days' work in one, plus also go and put the cattle into the railway wagons.'[35]

Harry Frank often used this gig when out droving in the Central North Island. The advent of the pneumatic tyre meant greater comfort when travelling with horse and gig.
Harry Frank Collection

bob-tailed horse and fairly heavy horse. And he looked round to me and said, 'You might float a bit here.' The rivers were up and running yellow, and we went down between the two streams and I thought it was alright for him because he was on Stumpy! He was – well we reckoned he'd been born on a wet stone in the middle of the river. And that was the end of the theory about Canterbury horses.[36]

Laurie was ready to admit, though, that horses bred on the high country or Canterbury Plains would still not be suitable.

The drover and his dogs

Tip, Old Joe, Tie, Sharp, Dune, Laddie, bloody mongrel – no matter the name, the importance of a dog in the working life of a drover cannot be underestimated. The drover was inevitably seen with a pack of four or five dogs on the road. As one drover put it, he doesn't know 'how many men it would take to replace a good dog'.[37]

Those brought up around stock learned the skills of working dogs more often than not by observation and trial and error, and guidance from other skilled stockmen. Many were just a child when they got their first dog – it might be a pup to train, or an older dog close to retirement. John Sullivan laughs as he remembers, 'You get a dog and after two or three days you learned to know what the dog wanted you to do, sort of thing.'[38] Laurie McVicar had a similar approach: 'The dog taught the boy more than the boy taught the dog.' He quickly added, 'A lot of those dogs didn't need working, they were just clever.'[39]

North Island drover Ian Mullooly remembered being taught to look after his dog:

> It was my first drove and I was given this dog called Red, he was an all-round dog, he'd head, hunt or lead, do anything. And Ernie [an older drover] always said to me, he said, 'You help your dog and he'll help you.' So, I was getting skinny and he was getting fat – you know, I had to run around as much as the dog was. Anyhow . . . we were going up the Otoko hill, you know how the road falls away somewhere and the fenceline down below and the sheep get down there grazing and you've got to put a dog down to come out. And Ernie always said get down and help your dog, so I used to go down and I was going down there with a dog and pushing them up. Anyhow this day I thought 'Oh, bugger this' – excuse my language . . . so I looked up the front to see if Dad was in sight, or looked in the back where old Ernie was sitting in the gig. He had an old gig with a thing underneath to carry the dogs. And I thought, gosh, I was only just bending down and talking to Red to bring these sheep up and I got a boot up the backside: 'Go down and help your dog.' I never ever forgot that. So I went down and helped the dog.[40]

The tried and true method of training the dog on a string or rope and teaching it to 'sit down' and 'come in behind' was the starting point. A variety of commands could be used: some whistle, some voice, some hand signals and the odd swearword thrown in. Some acknowledged with a degree of chagrin that the skill of whistling was beyond them, so they always used voice commands. Others resorted to the tin whistle, which was way more piercing, when stopping a dog.[41] Occasionally people took extreme measures to ensure they could whistle: Jack Curtis recalls his father rearranging his new set of false teeth with a file and hacksaw.

Perhaps swearing was a prerequisite for working dogs, if Harry Frank is to be believed. Harry, droving around Te Kauwhata area, recalls taking his wife out droving just after they were married. Bev was a horsey person, and loved being out – although she never had her own dogs, because 'she couldn't swear enough'.[42] Fred Cowin recalls:

> I used to claim that I could swear in seven different languages. You see the whole family of us at one time or another drove bullocks as draught animals . . . but you've got to talk to dogs, you've got to roar at them in emergencies and you don't know what you've said to them.[43]

The degree of competency drovers had in working their dogs varied greatly. Some were quick to admit they were no 'trialist', while others honed their skills on the dog-trial circuit. Bill Pullen, two-time national champion, even broke his dogs in on the road. And some droving characters are remembered for their dogs, even after their demise. Billy Riddle, a drover on the road from Gisborne to the Waikato, is remembered for his dogs being 'everywhere or somewhere else'.[44] A story is told of one of his dogs hiding in a large rural letterbox on the side of the road eating a big string of sausages.[45] At the other extreme, Arthur McRae could remember seeing 'jokers who could make a dog wag its tail with a whistle'.[46]

Robin Turner had 'special dogs for special jobs':

> Oh I had all sorts, I had heading dog, leader, side dog, huntaway. Got to have a bit of everything for the different things you've got to do. A side dog is when . . . if a car comes, this dog would come along and would move the stock over to let the car through . . . [and a] leader out the front so they couldn't get too far away. And I had another one that used to, if the sheep got all bunched up and wouldn't move, I had another dog that would go and jump on their backs and walk all over them . . . they didn't like that.[47]

The Drover's Dog

The old dog stands there, blinking in the sun,
Shaggy, blear-eyed; but years of work well-done
Have earned him right to bask, and eat, and doze,
In easeful warmth. And it may be he knows
That in this place he well may claim his own,
For this old track his busy feet have known
Since puppyhood. 'Twas here he learned his work;
To fear the master's hand that did not shirk
From rendering punishment when it was due;
To prize the words of praise though curt and few;
To heed the whistle – and to use his brains.
So now, in that short span that yet remains
Before death ends the old chap's useful life,
He stays at home, here with the drover's wife,
And sleeps, perhaps dreams that he is working still
Here where the stock-route winds across the hill.

—Kathleen Hawkins, from 'The Old Stock Route', 1942[48]

Harry Frank's ever-faithful dog team. *Harry Frank Collection*

> **TRAINING CATTLE DOGS**
>
> On this subject 'Komata' writes:— In answer to a letter addressed to the Editor on the above subject . . . I find a shy dog, who will not bear thrashing when in the wrong, as a rule seldom turns out well. If he comes of good stock, and is put to work with a good dog, he may do very well of himself. . . . Never on any account teach a sheep or cattle-dog any tricks, especially fetching sticks, nor to hunt cats, rats, or birds. Keep them to their work. If you want a dog to use for both cattle and sheep, break him in to sheep first; if not he will be too rough for them . . .[49]

A good dog was the envy of every drover. Skip Whittaker recalled a story told to him when he was 'just a kid', of relatives at Mangamuka trying to mould their dog into a winner.

> This couple there had an old dog called Bob – they used to watch the drovers go past time and time again and see how well their dogs worked. This joker . . . goes down and says to the Pakeha joker in the lead there, 'Jesus but you've got good dogs.' 'Oh, yeah, they're pretty well bred.' And the joker says, 'They've got pretty short tails too eh?' 'Yes, yeah that's the breed of the Smithfield, they're good dogs.'
>
> So he goes home and has a bit of conflab with his wife. And he says, 'You know, we can make this dog of ours good.' And she says, 'How's that?' And he says, 'Oh shorten up his tail.' So they decide to do this and get out there and the old chopping block, you know. And he says, 'Now, wahine, you back the dog up there and,' he says, 'I'll take the tail off.' And the dog was a bit fidgety and he comes down with the axe and says, 'Go away, Bob.' And Bob never ever came back. And he says to his wahine, he says, 'I think we cut 'im too short.'[50]

The early settlers mainly used border collies, brought in by Scottish shepherds who were working on pastoral runs. Border collies, huntaways, beardies, Smithfields, Kelpies, handy dogs and leading dogs were all working dogs that were advertised for sale in the newspapers of the late 1800s. Over time, new breeds and lines were introduced, and out of the mix came the huntaway and the New Zealand handy dog – both bred for the New Zealand farming culture.

The huntaway, the ultimate mutt and most commonly seen farm dog, was bred for performance rather than looks. It has its origins in the border collie, beardie, foxhound, Labrador and undefined other breeds.[51] The huntaway performs all manner of tasks, including heading and hunting. Their trademark bark is an essential element of their arsenal. Bill Pullen was much enamoured of the huntaway:

> If you've got a situation where it's a bit open or something you just want to gather your mob – you'll stick a huntaway out and the noise is applied and the stock will come together . . . I've worked with guys and they've only got one command . . . and the whole bloody lot go and attack. You're just causing trouble. You know you just want one dog – you put one dog out and standing there and make him bark the mob – if they're feeding [they] will up with the head and start walking and regrouping, and away you go.[52]

Others relied on the all-purpose New Zealand handy dog, which is believed to have come from a strain of huntaway and could do a bit of everything.[53] Noel Martin recalls that when he was out droving on his own, a good handy dog was essential:

> . . . then you wanted probably three dogs, specially one good handy dog that could jump – you could pop over a fence and hook . . . stock off little banks – a dog could get through a fence or jump well – just a handy dog – didn't have to be a great distance dog. He was one of your mainstays. In reality your cattle were drifting and feeding, the stock was drifting and feeding and there wasn't great dog pressure. You were under no pressure most of the time, occasionally you wanted a bit of noise. No, but it's not like mustering and or loading up into a woolshed or something where you want a lot of noise. It was mainly lighter work and handy dogs were the main criteria – you didn't want heeling dogs, . . . you just wanted nice quiet dogs. You were in fact treating your stock kindly and slowly and letting them drift and feed. And if you wanted to turn something back . . . you've certainly got to have handy dogs, not just big thumping huntaways – they've got to be dogs that can whip out with a little bit of pace and be handy, and as I say you can place them. If you've got a dog that you can place very well to pull cattle back off banks and that sort of things.[54]

The advantage of a handy dog was that they could run quietly but also bark on call; so they didn't disturb stock, but they added 'weight' when needed.[55]

While drovers might argue the toss as to breed, they were unanimous on who the real hero of the dog pack was: the leading dog. This was not a breed in itself but a strain of dog, often from heading dog stock, that seemed to have an innate ability to lead the mob from the front. A good leading dog was 'like another man' and 'the most important dog of the lot'.[56] They were used mostly by drovers, and are rarely found now.

The leading dog, as its name suggests, would be found at the lead of the mob, working pretty much on its own. It controlled the pace at which the stock moved – whether sheep or cattle.

> You could let your leading dog go and if the cattle were travelling too fast you just whistled him on, he would stop the mob, and you would give him another whistle on, he'd bring them back a wee bit so the back half sort of gradually kept up. Then you'd just tell him to get out of it and he'd just dance up the road in front of 'em.[57]

The leading dog stayed out the front:

> . . . all the time from when you let him out in the morning till when you put him away at night he's there all the time. And he just stays there, he won't come back . . . he's like a brake – slow things up – they know what to do – he can be out of sight –[he] could read the situation.[58]

Good leading dogs had an instinctive ability to recognise a previously used stock route and block gaps in fences, open gates and side roads. They would wait until the lead went past then race up to the front again. Some of the really good ones were even known to hold the lead up at the holding paddock at night:

> . . . you wouldn't even have to tell them because sometimes you're stretched out for maybe half a mile. So they'll just hold the bloody lead up, and then you get there and hello, here's the mob all held up and he's waiting there with them.[59]

Murray Dymock was quick to praise the ability of one of his leaders:

> Once she'd been over a road she knew every open gate. Didn't matter whether you went over it next week or next year, she had a memory

Allan Crawford and his droving crew snatch some lunch along the route to Paparoa in 1984. *Shaun Reilly*

A young Bill Pullen with hack and pack. *Bill and Gill Pullen Collection*

Sleeping rough was all part and parcel of being on the road. Bill Pullen relaxes beside his bivvy and fire while the cattle in the holding paddock look on. *Bill and Gill Pullen Collection*

Bill Pullen thought he was 'made' when he moved from sleeping rough to travelling with his truck – no more sleeping on the hard ground. *Bill and Gill Pullen Collection*

Bill Pullen and his mate Pat Coogan rest their mob on a drove through the Bay of Plenty, circa 1978. The caravan brought yet another level of comfort for the drover on the road. *Bill and Gill Pullen Collection*

The Munro family of Otematata still drove their sheep over the Lindis Pass to fresh grazing areas several times a year. A modern drove requires pilot vehicles in front of and behind stock, or warning signs placed appropriately. Some local councils may even require drovers to wear high visibility vests. *Lyndon Ferry*

Munro sheep on the move through the Lindis Pass. For today's motorists, coming across a mob of sheep being driven to new pastures is a novel experience. *Lyndon Ferry*

Barwood Motors of Fairlie has been operating as a cartage and livestock transporter for nearly 100 years. The stock trucks have come a long way in that time. This Isuzu has a four-deck sheep crate that can carry 350 ewes or 550 lambs. The crate can also be converted into a two-deck version, able to carry 42 two-year-old cattle or around 100 calves. *Susan Hayward*

for that road and she knew if there was a gate open – she'd be there, she'd just sneak away. You'd never see her go and she'd be right in the gateway. Never had to work her, she always did it herself.[60]

Drovers looked for certain qualities in a leading dog. Bill Pullen looked for a huntaway lead:

> Cattle and sheep in particular get really road-happy and they just march and march. They get ultra-fit. You know, I've seen, with big mobs of sheep, I've seen two of us in front of the mob, bloody just jumping up and down, dogs going and trying to stop them from going and they're just pushing us backwards down the road . . . they just want to keep going. So they take a bit of stopping – so an eye dog is not really much chop . . . but with a huntaway with his noise he can woof, woof, stick a bit of noise on.[61]

Bill also looked for a dog with a bit of gumption:

> A good dog, if a beast's running away and you want to stop it – the first thing they've got to do is grab it on the head, grab it on the nose or ear or somewhere or on the cheek, and they'll grab it on the nose and boy that makes them go up, and then when it turns they'll go in, woof, in along the heel. But you see a lot of dogs are too gutless to go in and take him by the head. They'll go and bite a sheep on the bloody arse but they won't bite a beast on the head. That's why heading dogs, as a rule, are a waste of bloody time on road jobs.[62]

Once a drover had found the right dog for the job, the relationship could last for years. Don Monk recalls going through three before getting one he was happy with, and then he worked that one for six or seven years. The first one would 'dance in the front' but would not 'pull or anything'; the second just 'didn't have it, if you know what I mean'; but the third one, 'the third one – he knew as much as I did'.[63]

An unfortunate reality of working out the front of the mob was that the lead dog was the first to meet oncoming traffic; and he often went unnoticed by traffic coming through a mob from behind. Bruce Ferguson, droving a mob of lambs for Dalgety's in Golden Bay, once encountered first-hand the lack of consideration for the lead dog.

> A chap in a car came up behind and he'd been there for a while and couldn't get through – you see, the sheep filled the road up.

> So I took pity on him and I said, 'Look, I'll send my dog alongside and he'll clear a track for you. Follow the dog through.' And that was alright, he went about from here to the back door [10 feet] through the sheep and then ran on top of my dog. So he lamed the dog too. Here's the old dog walking along wagging his tail going woof, woof and this bloke behind next thing he rides up on top of him. So I yelled out to the bloke who had the next cut, I said, 'Keep that so-and-so in the mob, don't let him out.' And I'd have kept him there for the rest of the day because he'd run on top of my dog. But anyway this silly bloke who had the next cut, he let him through. Nothing I could do about it. I couldn't do anything about it.[64]

The loss of a good lead was a terrible blow to a drover; they were not always easy to replace. Insurance was no salve for the loss either, as Spencer Dillon recalled when he lost a leading dog in the Wairarapa.

> Taking cattle down to the rail one morning I had quite a good little leading dog and we went . . . over the main road and straight in behind the trucking yards. And these cattle . . . went down to the right a bit and I called the leading dog on and he turned the cattle coming back, and he just sort of walked to the side a bit to look back up – a car come along – boof – hit his head – and they hadn't slowed down, never stopped – that was him. Anyhow I'd taken out an insurance on dogs and rung up the insurance company and told 'em. Anyhow a week or two went by and they didn't want to pay up, and I said, 'Well, I'm being honest with you, if I'd told you lies you would have paid me by now.' Anyhow this joker said, 'Are you going to keep your dogs insured?' and I said yes so he sent me a cheque and I never kept the dogs insured.[65]

The way drovers cared for their dogs differed greatly. At one extreme there was Davey Gunn: reportedly 'nothing was too good' for his dogs. When he was eating his porridge, 'he would take one spoonful for himself, then give a dog a full spoon, then take another spoonful himself'.[66] Others were perhaps not so particular. According to Lee Fyers, Billy Riddle's dogs were not well fed until possums were frequently found as roadkill. 'Like he had rough horses and thin dogs – until the possums came and when the possums were run over on the road – well the dogs were in good order.'[67]

Dogs were always well fed at the drovers' whare on Fairview Station, a Lands and Survey block in the Bay of Plenty. As Jack Curtis remembers it there was:

> ... endless dog tucker ... The old man bought horses ... from the pound in Taneatua – well everybody called them Maori horses but they're not really Maori horses, they're just horses – shot them by the dozen ... sometimes one every day. You'd get five drovers pull up at home and the old man would say, 'How you off for dog tucker?' and he'd go down with a rifle and shoot a horse, skin it and pull it up the big willow tree and then go down tomorrow and it's just a skeleton hanging there.[68]

Bruce Ferguson recalls a story about old Jack Flower's dog fending for itself.

> Jack Flower ... had a good leading dog – he loaned the dog to Teddy Rosser and Rosser can't have been feeding him enough. Jack told him, 'I warn you, Rosser, if you don't feed Dandy he'll feed himself.' I don't know who Rosser had driving, but he had someone bringing his sheep up and he'd ridden up ahead to one of the farmers, he was trying to make a deal and buy a few ... so he's inside having a cup of tea being nice and so on, and when he went out Dandy had killed and partly eaten the pet lamb on the veranda.[69]

One old lady in Rotorua had her fish pond just about drained by thirsty dogs. Jack Curtis still remembers the bollocking he received from the irate woman – 'Well, did she go to market.' That same drove, the kindness of two women further down the track was greatly appreciated: 'Here's two old girls with a bucket each full of water for the dogs – "Oh we feel sorry for the dogs, we feel sorry for the dogs." '[70]

The SPCA might be onto a drover if there were concerns about the care of his dogs. Guy Shanks referred to Charlie Aiken, the SPCA inspector, as a mongrel, after getting into trouble over the care of his dogs.

> When I was living at Ruawai beside the saleyards, old women used to ring him up and tell him I hadn't been home for four days and my dogs hadn't been fed. I'd feed them before I went and my mate would come and feed them. I had about 10 dogs and I didn't take them all. You need fresh dogs for the next trip. There were a couple of old tarts down the road – they'd ring Sid Wells the policeman, too. I was never prosecuted but I got a letter from Aiken.[71]

Drovers often had as many as 10 dogs, divided into two teams: while one team worked the others had a rest. For drovers like Keith Sowerby and Charlie Sundgren who were droving seven days a week, it was important

Ken Lewis with his trusty companions on Ninety Mile Beach, circa 1981.
Northern Advocate/*Mike Hunter*

to be able to rest their dogs. When the dogs were footsore, particularly when they got back on the road after a prolonged rest, Jack Curtis swore by a mixture of pork lard and 'stockallum' tar smeared on their paws.

Dogs were faithful companions on what could be a lonely job, and their loss was often taken hard. Kath Burton recalls the only time she saw her father cry was when one of his favourite dogs died of distemper.[72] Owen Gibbens, too, recalls the loss felt by his father when three of his dogs were killed by a shunting train in Greymouth: 'if you've ever seen a broken man'.[73] Drovers had no compunction, however, about ridding themselves of any dog that failed the grade. Bill Traves remembers his father having good dogs but comments, 'if they were no good they didn't last long!'[74]

Much like fishermen's tales the stories of drovers' dogs tended to grow in the telling. The story of Dave Beazley's obedient dog is perhaps one such tale. While taking sheep to the Moerewa freezing works, Dave Beazley of Northland loaded them onto a pontoon to take them across to Mangungu. Some of the sheep unfortunately broke the rail and went overboard into about 12 foot of water. Dave 'put his dog around the sheep, which were swimming quite well and making progress back to the pontoons, when he noticed one tiring. The only thing to do when a sheep is tired is to take the pressure off it, so Dave whistled his dog to sit down – and he hasn't seen it since!'[75]

There is always a certain tone of reverence when a drover talks of his dogs. Hugh Monahan philosophises, 'It's a matter of how good a man you are with a dog, what it rounds up to.'[76] Jack Curtis puts it more baldly: 'without 'em you're buggered'.[77] Perhaps, as Vernon Wright comments in *Stockman Country*, it is all to do with their inextricable relationship with the canine.

> At night in the endless tales around the fire, the man becomes his dogs and the dogs are the man. When a stockman launches into a long and impassioned story about or involving his dogs, even a bush psychologist can tell he is actually talking about himself. The device is a good one and some might call it very healthy. Thus in the sometimes almost neurotic undemonstrativeness of New Zealand male society the stockman is able to talk about himself, his day and its successes and failures in emotional terms, discussing his dogs as though they were factors quite independent of himself. In the man's emotional life the two are almost indivisible, even though every stockman may deny it to his last beer.[78]

Both the horse and the dog were imperative to the drovers' work. The drovers were proud of their four-legged partners and, while they were out on the road, unsure of what any day might throw at them, they certainly came to rely on them.

Sheep trucks parked outside the business of Graham & Gebbie in Hastings, circa 1930s.
Alexander Turnbull Library, 1/1-019209-F

8. TOOLS OF THE TRADE

The necessities for survival on the road

As a rule, to drive a mob of sheep for any distance, takes not less than four hands – with three horses saddled, and one packed with provisions and baggage – but . . . when one says 'baggage and provisions', these are to be taken in a limited and qualified sense. A pair of blankets, and a tin quart-pot per man; a frying-pan, tether ropes, butcher's knife, a bag of flour, and canvas bags of sugar and tea; a knob of salt, and a few boxes of matches – with a reserve of flint and tinder for a pinch – in most cases make up the list . . .[1]

While the list of necessities opposite may have been a little out of date for those on the road in the greater part of the twentieth century, the planning still had to be done. Some drovers were immaculately turned out and fastidious about their gear; others were a little more ad hoc. Food was usually just 'knocked together' as quickly as possible. When it came to their dress, practicality definitely won over fashion. And there was a degree of pragmatism, too, about where they spent the night.

Clothing

Drovers wore clothes that would withstand the rigours of their work and protect them in the extremes of weather: there was 'nothing exotic about the clothing'.[2] In the early years of settlement it was hobnail boots, heavy work trousers — often of woollen cord or moleskin — and a blue serge shirt. This last was a great leveller: men of all backgrounds wore it, and Charles Tripp of Orari Station even recommended it when he was back 'home' lecturing on life in the colonies.

> . . . I recommend woollen clothing of the strongest description, and such clothes only as you would wear in the country here. I have known many people with very limited means encumber themselves with a lot of fine black cloth clothes, whereas if the same amount had been expended in good Fustian, or woollen cord trousers, they would have been better off in the colony. Again, instead of coats being worn up the country, we wear a blue serge shirt . . . a good supply of very strong nailed boots, . . . I have found most valuable. Most of my black clothes and thin boots I took out with me about nine years since, I have brought back, and I have hardly worn them a day since I left — except in the towns.[3]

William Swainson tells an anecdote about Charles Bidwill's cabbage tree hat.

> . . . it must have been at the end of April, 1844, when your father and myself started from the Hutt with his sheep . . . The universal dress in those days for country settlers and travellers was moleskin trousers, blue serge shirt, and that much-prized head covering, when it could be obtained, a New South Wales cabbage tree hat with a long black ribbon. The older the hat the more it was valued. Indeed I have known a new one willingly exchanged for a much older article – perhaps because it did not look so 'new chummy.' They were difficult to procure, having to be ordered from Sydney, and the price was high, from 25 shillings to two pounds, but the time they lasted was surprising . . . I never shall forget your father's hat blowing off on a piece of sandy beach near the Mouka Mouka Rocks. Away it went and your father in pursuit, with a bundle on his back and his gun in his hand. As soon as he was near enough to induce him to stop to pick it

A group of drovers having a well-deserved drink after a drove to Whataroa saleyards in 1950. From left to right: Eddie Nolan, Steve Nolan (front), Lewis Condon (back), Kevin Nolan (with cup), Tommy Condon (front), Mickey Clark. Their clothes are typical of a drover's clothes for the time – including the ever-present oilskin. *Mickey Clark Collection*

up, put his foot on it or stick the muzzle of his gun in it, it would elude him and start off at a racing pace, till at last a wave caught it, filled it with sand and it lay quiescent. Your father picked it up and being pretty well exhausted sat down, looked at it and said something. I did not hear what, but can imagine. I don't think I ever laughed so much in all my life and that did not improve matters.4

The drover's list of clothing might include: a woollen singlet, woollen longjohns in winter, a shirt (often woollen), woollen socks, a woollen jersey or a Swannie, a woollen sports coat, an oilskin coat and leggings, a sou'wester, underpants, good strong workboots, and one set of tidy clothes (optional). Laurie McVicar reckons they never went anywhere without an oilskin coat – as the saying went, 'If it's fine take a coat, if it's raining please yourself.'5 Betty Eggeling recalls always taking 'good clothes', packed in an oilskin on the packhorse, for nights when they stayed at someone's home or at the Fox Glacier Hotel.6 And Lee Fyers remembers one tidy old drover who would come out in the evening dressed in a jacket and bowtie.7

Some drovers gained a reputation for their appearance – good or bad. People noticed the difference between two drovers often seen at the Waiwakaiho saleyards in Taranaki. There was a stockman named Johnson who, despite having been right into the backblocks to bring out stock, would appear at the saleyards fastidiously dressed and with his horse looking as if it had just 'finished a session under the brushes of a show-stables groom'. Then there was Tommy Hook, who, with his unkempt appearance, was always dealt with out in the open air.8

The oilskin coat, leggings and sou'wester were the most commonly worn garments – particularly for any drover working the west coast of the country. Tony Condon remembers the advice of Johnny Heveldt, an old-timer in charge of the packhorses on the iconic South Westland drove: 'Any silly bugger can wear a coat on a wet day, bloody good man that takes his coat on a fine day. Tie it in front of your saddle.'9 With not a cloud in the sky when they set off, Tony had his doubts, but as the rain came down mid-afternoon on the Maori Saddle he was grateful for Johnny's wisdom.

The drover's personal grooming was basic:

> We'd just have a scrub in a trough or a creek you know – you had to wash regularly or with the hot weather with your woollen clothes on and that you'd chaff. You had to keep your hygiene up. I'd use to shave every day, always shaved, never missed a day without shaving – with cold water in the dark, with a blunt razor.10

While many drovers could recount how they washed their clothes in rivers along the droving route, not many could say they washed nappies. Gill Pullen used streams to do her family's wash when out droving with her husband Bill and their young boys, Roy and Ian. Here she is doing her wash between Hicks Bay and Cape Runaway. *Bill and Gill Pullen Collection*

As for laundry, clothes were left under a rock in a creek and the current or a waterfall did the work of agitator. The clothes were dried in front of a fire at night in a hut, or, when roughing it, alternative methods were used: 'It was nothing, absolutely nothing to see the old packhorse wandering along the road with a pair of pants or pair of underpants hooked on the side of the saddle – no problem.'[11] Of course, if they were staying the night in a pub or with friends, they would take advantage of more modern facilities, as Bill Pullen recounts:

> One time I was on the road with this chap. And I was in the lead this day, which was unusual for me – usually I was behind the mob. Anyway this lady came down the driveway and we were actually stopping the night there, right at the paddock there right at her house. And she said to me, 'I'm just slipping down the road to pick the children up', she had about five or six mile to go or something to pick her children up from the school bus. So, she said, 'If you'd like to shoot in you can have a shower or bath or whatever it was while

you're waiting for the tail to come up.' So I left my leading dog out
the front and tied my horse up on the thing – well I was in there like
a rocket. Any rate, my mate comes along and I bloody lean out the
bloody window with the towel, waving it, 'Do you want a bath?' He
couldn't believe it. He couldn't believe it eh. Then I of course came
out on the road and he said 'Jeeesus,' he said, 'what are you doing in
there?' I said, 'Having a bath, mate.' He said, 'Is there no one home?'
I said, 'No.' So that's the story I tell, but I don't tell 'em that the lady
said I could go in.[12]

A well-known droving identity from Wairoa explained why he watched his appearance: 'I think that's one of the reasons why I shaved every day, you know what I mean? If a scruffy-looking bloke came in and asked you for a paddock, you wouldn't get it. But I'd come in – I'd be clean and tidy and they'd talk to ya . . .'[13]

Droving's culinary delights

Diary accounts of early droves describe loading packhorses or drays with essential items such as flour, sugar and tea; and later, tinned meat or fish were added to the mix. Their supplies were by no means luxurious: many entries speak of minimal rations and a certain resignation after a hard day.

> We all met at the tent at the same time with glum faces . . . having faced an unsuccessful day of searching for stock, fryed the mutton I had brought, ate some bread, the only thing we had had for the day except one of my boxes of sardians [sic] and retired to our blankets.[14]

Rations could be augmented by the judicious use of a gun.[15]

In these pioneering days, drovers who were running close to the bone were often helped by station owners or fellow drovers.[16] As the country became more settled, accommodation houses were built at regular intervals and meals could be bought – although the food there often had its limitations. John McGregor, on a drove through to the Mackenzie Country, wrote:

> The Landlady Mrs Waldin when we were at tea came to the door & put her head through with as much expression as possible, 'mind-mind' there's a pudding coming. The meaning of this was not to consume but as little meat as possible, meat was dear in those days.[17]

In the twentieth century, travelling with a packhorse, the ubiquitous Cooper's dip box was often the drover's favoured tucker box – no matter that one of the more toxic ingredients of the original contents had been arsenic! Another variant was the butterbox from the dairy factory.[18] Some or all of the following would invariably be stowed away inside: spuds, onions, cabbage, carrots, bacon, sausages, condensed milk or sugar, bread, often 'green bread till the next shop',[19] butter and jam – although some never bothered with such niceties as these last items: 'You never bothered about jam or anything like that, oh Christ no.'[20] Arthur McRae remembers a story he heard of old Ken Harding going through the Motu (Motu Gorge) with a new recruit.

> He said he got up in the morning – you know, a young fella on his first job – he got up and got the fire going and the billy on, cooking toast on the embers and that and put on his butter and jam on his toast and Ken, he said, 'Butter or jam . . . It's a long way to the next store.'[21]

Some tucker boxes also held tins of bullybeef and tinned peas, later replaced by dehydrated peas. Plum duff or fruit cake was a good filler, with the added advantage that it didn't squash too easily.

Three sizes of billy, each smaller than the other and packed inside one another, were also packed in the tucker box: 'three billies, and a sugar bag so it wouldn't rattle . . . and frighten the horses . . . three billies, big billy for boiling up the water and then the spud one and the tea billy was the smaller one'.[22] Some even had a butter billy – a small white enamel-lidded billy, inside the last of the billies, to stow away a pound of butter. A camp oven (a cast-iron pot with a lid) was loaded up on the packhorse, along with fire irons and a metal tripod to hang the billy on over the fire. Those with the skills could rattle up scones in the oven; and Jack Curtis's father even made sponges in them.[23]

In the 'modern' era, Road Service bus drivers and truckies often obligingly acted as couriers for drovers, dropping extra supplies off to them along their routes. At the occasional country stores, drovers might purchase a sweet treat such as a bar of chocolate; it would be downed 'like a glass of water – you're eating tucker that's so bland and basic like spuds, boiled meat, cabbage if you're lucky . . . if you get on to something sweet imagine . . . I mean, just whoofo it's gone.'[24] Sometimes food came unbidden. Local Maori on the East Coast would supply drovers with fresh fish and locally grown vegetables;[25] and around the country, kind womenfolk would provide cups of tea and sometimes scones so the drover could stand at the farm gate and have a cuppa while they watched the mob slowly meander past. Ray

Stevens recalls the kindness of one woman when he was moving a dairy herd for a sharemilker:

> . . . going over the Kaimais – that's a big hill climb from the last of Waikato – when you get to the top you're in the Bay of Plenty. I can remember the rain on the caravan roof as we went to bed that night and it was still there the next morning. So it looked like a wet day . . . but we knew where our night paddock was. From the house, the lady of the house must have seen us enter, cos by the time we'd fed the dogs and got in and got out of wet clothes the lady arrived and lifted the boot of the car and brought over a Pyrex dish full of chops and onions and spuds. And those are the type of people that we met. And it was so wonderful. She assured us that she'd been thinking of us all bloody day, and the boys would like something warm and there it was. We didn't ask for it but it was certainly appreciated.[26]

Left: An advertisement from *New Zealand Farmer* magazine, 25 November 1960. *Hocken Collections, Uare Taoka o Hakena, University of Otago, S13-377a* Right: Keith Sowerby takes a moment to pause and relax over a cuppa during a drove in the Manawatu in the 1970s. By that stage there was no need to boil the billy; a thermos flask of hot tea provided instant warmth on a cold day. *Julia Sowerby Collection*

DROVERS'

CHARLES TRIPP'S DAMPER
Flour
Water
Campfire

Mix ingredients as you would bread. Scrape away most of the ash from the fire. Put dough there and cover over with ashes. Test in an hour by putting stick/fork/knife through it. If dough does not adhere it is baked.[27]

FRANK LAMBERT'S PINK POTATOES
Saveloys
Potatoes
Water

Place potatoes, saveloys and enough water to cover in billy. Place billy over campfire and boil till cooked. You will find potatoes go a lovely shade of pink from the saveloys.[28]

BILL PULLEN'S 'FLASH' DINNER
Mutton or sausages
Cabbage
Spuds

Using your camp oven with the bloody legs hacksawn off, place over campfire.

Prepare meat, spuds and cabbage. Place in camp oven at intervals appropriate to cooking times.

Then you cook shit out of it till five minutes' time you're ready for a meal.[29]

HUGH MONAHAN'S EGGS AND TEA
½ dozen eggs
Kettle
Water
Tea

Pop eggs into kettle with water. Bring to boil. Take eggs from kettle. Use water to make the tea. Take cup of tea to Aunty. Ensure Aunty does not find out your method so as not to get your pedigree read.[30]

MURRAY DYMOCK'S SPAGHETTI/BAKED BEANS SURPRISE
This is a very convenient meal on the go!
Can of spaghetti or baked beans

Open can with butcher's knife as you are walking with the mob – horse following along behind. Use butcher's knife as eating utensil. The surprise flavouring is whatever is on the knife from its last use

COOKBOOK

(it may have been used to slaughter an injured beast).[31]

IAN MULLOOLY'S PLUM PUDDING
Tinned plum pudding
Little empty bottle

Take plum pudding out of tin. Take little empty bottle and wait until farmer has dropped off cream can at gate. Fill little bottle with cream – without getting caught! Add cream to plum pudding.[32]

ROBIN TURNER'S FRUIT SALAD
Green apples
Hard peach
Plums

Each drove, recall where the local orchards or fruit trees are en route. Pick and eat at your leisure.[33]

MAX DOWELL'S CRAYFISH TAILS
Crayfish tails

Go out cray fishing before the drove. Cook crays. Pop cray tails in coat pocket – eat at your leisure while riding behind your cut.

MURRAY DYMOCK'S TEA
Hot tea
1 bottle
One sock or two

Make tea. Pour tea into bottle. Place bottle inside one or two socks – depending on how long you want it to stay warm. Drink tea at smoko – bound to be cold.[34]

JIMMY DE-ARTH'S BLACKBERRY SANDWICH
Bread
Butter
Blackberry jam

Cut bread into 2-inch slabs. Slosh on a great deal of butter. Add huge lashings of blackberry jam. Clamp pieces of bread together. Wrap in *The Daily News*. Place in saddlebag for lunch.[35]

PETER EGGELING'S BILLY TEA
Billy
Water
Tea

Just boil water over open fire. Throw in handful of tea. Put lid on billy. Tap billy with stick to make tea sink to bottom of billy. Drink when ready.[36]

Don Beard recalls, as a child, watching with disbelief as Charlie Sundgren downed hot tea in seconds.

> [We'd] hear him coming a mile away – get the billy on sort of thing. He was a really hard case. I always remember one day a real raw winter day up there, 700 feet, and he must have been like an iceberg and he came in, can't remember how he got his wet weather gear off cos his fingers must have been numb, but he took this cup of boiling tea and drank it straight down. Even Mum stood back and gasped.[37]

As a young child, Jessie Bradley (née White) of Buller, was always thrilled when she saw drovers heading towards their farm. She loved to see their dogs, but better yet was the hope of receiving a coin as payment for a cup of tea.[38]

Mutton was another staple for drovers: a knocked-up sheep would become tucker for drover and dogs alike; and those droving cattle would always have half a dozen rams in the mob as a ready source of dog tucker as well as fresh meat for the drover. Of course, the best cuts went to the drover.

Boiling the billy was synonymous with a moment of rest and refreshment. There is no more romantic an image than that of a man relaxing with the billy boiling while the stock rest alongside. Like many such images, though, sunshine is a key ingredient; and the weather was not always so obliging:

> It could get tough on the bloody body because, for example, in the winter time as you can imagine sometimes it might rain for a couple of weeks – so – there's no breakfast. No such thing as breakfast – by about midday you might dive in – you know, get up under the pack cover, put your hand in the tucker box, feel around and pull out a spud or pull out a sausage, uncooked, and gradually find a bloody hunk of bread or something and you just devour it without butter or jam or any damn thing.[39]

If they were stuck for firewood on wet days, they might resort to the old trick of splitting splinters off totara fenceposts along the route, to be used later in the day.[40]

There was no shortage of water in streams along the routes.

> When you were going through the gorge it was beautiful pure water coming off the hills there, and you just went and got a cup . . . a lot of them had cups beside them – jokers left cups there so you could have a drink . . . no problems; dogs had plenty of water too. [41]

And a break at a country pub along the route was known to happen on the odd occasion – to the consternation of the local constabulary, if stock were found untended.[42]

Just a decade ago, meal arrangements were truly civilised in Northland. Ben and Christina Morunga, out on a three-day drove, organised each lunch stop: the first was at the restaurant at Panguru because a 'good feed'[43] was assured; the next day they could stop in at their home; and on the last day 'you can go past the shop and get a feed there or else you can get somebody to go ahead and come back with some pies or something . . .'[44] No Cooper's dip box in sight.

Considering the long, hard days drovers faced, they can seem quite offhand about something as important as food. Cooking was basic at best, and it was often a matter of making the most of what they had. While breakfast might extend to porridge and treacle or bacon and eggs, it could also be 'what we had for tea. Savs, sausages, spuds all boiled up . . . red potato, pink potato, yes – some would make their bloody tea with it – oh, shocking stuff. One boil-up for everything.'[45] No matter – as Jack Curtis explained, 'we certainly didn't starve [though] there was the odd time that we went hungry'.[46]

Accommodation

As essential as good food and suitable clothing was the ability to enjoy a good night's sleep while out droving. In an uninhabited landscape early drovers had little choice but to 'sleep rough'. Hardy souls had little option but to sleep out in the open air, or in a two-man pup tent, or under the shelter of a dray. A pair of wool blankets was included in the drover's list of basics. To soften the ground, vegetation was gathered and laid underneath the waterproof groundsheet. Edgar Jones, who travelled over to the West Coast in the 1860s with stock, describes his sleeping arrangements: 'At night time, we broke off short lengths of Birch tree branches, which we laid on top of one another for "feathers" on the ground inside the tent, and then the waterproof ground sheet, and our blankets.'[47] Samuel Butler, one of the early successful Canterbury pastoralists, captured both the beauty and the hardship of sleeping out in the open:

> On one of these flats, just on the edge of the bush, and at the very foot of the mountain, we lit a fire as soon as it was dusk, and tethering our horses, boiled our tea and supped. The night was warm and quiet, the silence only interrupted by the occasional sharp cry of a wood-hen and

the rushing of the river, whilst the ruddy glow of the fire, the sombre forest, and the immediate foreground of our saddles and blankets, formed a picture to me entirely new and rather impressive . . .

Our saddles were our pillows and we strapped our blankets round us by saddle-straps, and my companion (I believe) slept very soundly; for my part the scene was altogether too novel to allow me to sleep. I kept looking up and seeing the stars just as I was going off to sleep, and that woke me again; I had also under-estimated the amount of blankets which I should require, and it was not long before the romance of the situation wore off, and a rather chilly reality occupied its place; moreover, the flat was stony, and I was not knowing enough to have selected a spot which gave a hollow for the hip-bone . . . Early the next morning the birds began to sing beautifully, and the day being thus heralded, I got up, lit the fire, and set the pannikins on to boil; we then had breakfast, and broke camp.[48]

George White, prospecting for a suitable pastoral run in the Wairau, encouraged others to learn from his experiences:

One tarpaulin for a camp, to be erected on four poles; two in front four feet high from the ground and two eight feet back, eighteen inches to two feet longer, with four crosspoles each way on the top, the width according to the number of the party; beneath which you may sleep in spite of wind and weather as snug as a 'bug in a rug' . . . One thing more; sew up the bottom and half the side of your blanket to prevent your toes from looking out for the sun before your nose.[49]

As the European population increased and land routes between settlements became more defined, entrepreneurs seized opportunities to open up accommodation houses at key sites, particularly at river crossings and a day's travel from any other accommodation. One requirement of a licence was to provide holding yards for stock and, if at a river, a ferry to transport travellers across.

Conditions in the houses were by no means luxurious – often a space on the floor in front of a fire was all there was to be had, and where beds were provided the bedding was not always clean. One traveller wrote in disgust at the condition of some accommodation houses: 'I myself have known persons to sleep out under a flax bush rather than go into these places, where the beds were absolutely filthy, and where the promotion of drunkenness was the normal occupation of the publican.'[50] Of course, many houses were efficiently run, providing comfort and relief for the weary

Tophouse was built from cob in 1887 and still stands today. It is located on a low pass between the Wairau, Motupiko and Buller river valleys. The hotel became an important resting point for drovers travelling with stock from Nelson through to Canterbury. Earlier accommodation houses were built close to this site, one in 1844 and another in 1852. *Nelson Historical Society collection, Nelson Provincial Museum, 326773*

traveller. One wonders, though, how the drovers coped at Helen Gibb's accommodation at Motunau on the road to Cheviot when prayers and Bible readings preceded breakfast and dinner![51]

Along regular stock routes where there was no suitable accommodation, the odd hut might be knocked up; these were very basic tin sheds that provided shelter and little other comfort. Maud Moreland, travelling through the South Westland in the early 1900s, came across one such hut, irregularly used by cattle drovers at the Blue (Moeraki) River, and describes its starkness.

> But that night it was dark; it was wet; we were both tired and hungry and longing for a decent resting-place, and when we pushed open the door of that hut what did we find? It was nothing but a corrugated-iron box, eight or nine feet square, with a rude bunk, covered with fern, at either side; between them was a dirty cupboard smeared with candle grease, which served as a table; a stool by the wide hearth and two old billies completed the furniture. An axe-head lay near, but the handle had been burnt for fire-wood, and the floor was littered with dirty paper, old tins, sticks, and ferns. A more truly uninviting place would be hard to find . . .[52]

Staying at the Fox Glacier Hotel meant a night of comfort for the 'Far Downers' on their drove to Whataroa. Betty Eggeling would always wrap some tidy clothes in her oilskin so that she could ensure she was appropriately dressed when staying there. *Alexander Turnbull Library, WA-53608-F*

Private accommodation was also often provided by family and friends, or folks along the stock routes who were keen to make a little extra on the side. It took all sorts to provide such respite; some were held in the highest esteem and others gained a certain notoriety. Those droving stock from South Westland up to the Ross railhead may well have stayed at Sarah Bergman's at Kakapotahi. Mrs Bergman was held in the highest regard; John Sullivan described her hospitality when droving stock up to Ross:

> . . . She had a heart of gold, she was always cooking food for somebody . . . She was always jovial, always willing, knew lots of things about lots of people . . . she had a good telegraph system. She was kindly and good-natured . . . always speaking good of anybody.[53]

Where drovers congregated, yarn-telling over a drink in the evening was inevitable; and for children listening in, there was an air of magic about it all. Laurie McVicar's parents ran the Totara Flat Hotel, where drovers would often stay. For Laurie as a young lad:

> We were allowed to go and sit in the bar parlour of an evening . . .
> and people would come in . . . I suppose in a sense as children we
> shouldn't have been in the area – but we spent a lot of time in there....
> And there were quite a few old fellas that wouldn't ruin a good story
> for the sake of the truth and we found it, well I did, found it very
> interesting. As I say my role models were stockmen.[54]

Carol MacKenzie's family home in North Taranaki offered accommodation to drovers. For Carol the experience of being able to listen to their stories is something that she has carried all her life: '. . . the evenings when everybody's sitting around laughing and they'd be telling stories and nobody would want to go to bed. They'd tell us about their horses and their dogs and people – you know. It was like a book really.'[55] Minnie Grieve's accommodation on Banks Peninsula was remembered, however, for quite different reasons.

> There was an old girl at Devil's Knob, she had the holding paddock
> and she would accommodate the drovers and she had a bit of a bloody
> menagerie – she had roosters and possums and rabbits and all sorts –
> in the house! Jimmy Watson stayed there the night I did . . . She told
> me that 'Jimmy Watson must have hens, has he?' I said, 'Oh, I don't
> know – I don't know Jim that well,' and she said, 'He offered me a
> pound for that rooster.' This is going round in her old head, you know.

John McLennan stands outside the Maruia Hotel with his cattle in the late 1930s. Country hotels often had holding paddocks for stock to rest in overnight. For the drover, the hotel provided comfort and an opportunity to catch up with others over a beer. *Margaret Lusty Collection*

> And I met Jim Watson the next day . . . and I said, 'What the hell did you offer a pound for that rooster?' He said, 'Worth a pound to get a good night's sleep there,' he said, 'I would have rung its bloody neck.'[56]

Minnie was also known to keep a close eye on the number of stock in the holding paddock for fear of losing income. While other accommodation may well have been charging just a penny a head for the holding paddock, Minnie charged tuppence and she was not to be done out of any of it.

> . . . and she made sure she knew how many cattle went past. Course being first time down there, she said, 'If you'd like to let your cattle out,' she said, 'just open the gate and put 'em out on the road and head 'em this way, and let 'em feed along the road.' I thought, 'That's good.' So that's what I did – I opened the gate and let the cattle out on the road and went back to the house for breakfast and walked inside just in time to see her leave the window. She'd been there counting them as they walked past. I said, 'Seventy-one,' I think I had. She said, 'I make seventy-two.' She could argue the count; for the sake of a penny, I think I paid her.[57]

Accommodation could also take the shape of shearer's quarters, hayshed, pigsty or fowlhouse[58] – and sleeping out under the stars was always an option. The pack cover protecting gear on the packhorse was indispensable here. Bill Traves recalls his father Harold taking a 6 x 8 foot tent with a ridgepole when away on droves. If the good old Canterbury nor'wester struck, Harold would tie the tent ropes around underneath and lie on them so the tent could not blow away.[59]

There were no luxuries – no sheets, just woollen blankets tucked in the bedroll and, later, perhaps a sleeping bag; and a spare change of clothes acted as a pillow tucked into the hood of the bedroll. One of Jack Curtis's mates, who was smaller than most, just used a chaff sack as his bedding – and if it was cold, he used two. Arthur McRae remembers sleeping out, 'rain, hail or shine'. Despite tough conditions, he never remembers getting sick: 'If you got wet and your clothes got wet you'd just get into bed . . . and you'd be dry by morning. Never got a cold, never got the flu, never ever got sick like that on the road.'[60]

In the late 1960s, life for drovers on long hauls was revolutionised by the purchase of a truck – often an old Bedford. With the simple addition of a canopy on the back, they had a home away from home – and life on the road became easier. Still, it was by no means luxurious. Bill Pullen recalled his first Bedford, purchased for £1200 – the money was from the sale of

The Drovers Memorial Huts are of special significance to the Hawke's Bay. Built after World War Two, they commemorate the lives of three past Hawke's Bay drovers killed at war. A hut was built near State Highway 50 at Maraekakaho in memory of Lance Sergeant Wilfred James (Togo) Kirkley of the 2nd New Zealand Divisional Artillery, 5th Field Regiment, who was killed as a prisoner of war on 9 May 1945, aged just 27. Trooper James Edward (Jack) Oliver served in Egypt in a New Zealand cavalry division and was killed in action on 23 January 1942, aged 27; a hut was built in his memory in a paddock near Middle Road in Havelock North. The third was built next to the Taihape road, just below Pukehamoamoa School, for Private Percy Botherway, who served in the New Zealand Infantry, 24th Battalion; he was killed overseas at the age of 36 and was buried in the Sangro River War Cemetery in Italy.[61]

The Drovers Memorial Hut at Pukehamoamoa, which was erected in memory of Percy Botherway. *Hastings District Council Records*

1200 possum skins at £1 a piece. The canopy was a stock crate with a large canvas over it. Bill knocked together a fold-down flap, held together by a broken bridle, on which he could do his cooking on the white spirit stove – 'modern technology'[62] – as well as fold-down canvas cots.

> God it was so good. If it was raining it was home sweet home, it was so good. I mean people would look at it and say, 'God, how bloody rough and primitive.' And: 'What a shocking camp.' But boy was it better than lying out in the rain. It was so much better. And you had water! And you had food! And you had a stove you could cook on . . . You'd

be on the road and you'd be moving along and it come to midday and
you'd just . . . whip back and you'd get the old stove going and boil the
billy and you could actually have a nice hot cup of tea and you know
a decent sandwich or two . . . And at night time you know you could
cook a reasonable meal and yeah you had a BED! Even though it
never had springs or wasn't any mattress, you're just laying on canvas
– it was hard as hell – but boy was it better than on the bloody ground
month after month after year after year.[63]

Of course, there were other advantages to having a truck: surplus gear could be carried, such as wire strainers and staples in case a holding paddock fence needed a patch;[64] a 'half-pie horse float' could be towed; and if a beast 'knocked up' it could be transported easily; it could house the dogs;[65] and a horse could be transported in the back of the truck after a drove was completed.[66] And perhaps most importantly, you could 'get in the truck and go and have a beer, you know, after work if there was a pub handy, or go to the shop and get fresh stores . . .'[67] Ray Stevens, though, was less enamoured with the benefits of the truck:

> Ken . . . was a bugger of a man for getting somewhere central and he
> would do two or three days each side, and at the end of the day it was
> travelling back at the end of the day to the quarters we were to live in
> for four or six days. And at the end of the day the last thing I wanted
> was three quarters of an hour or so in the back of the bloody truck
> and again in the morning. Maybe I'd got spoilt by the dairy farmer
> having his caravan and our night's accommodation being where the
> cattle was. That was one aspect I didn't enjoy.[68]

Not long after the truck came the caravans – or even a campervan, in one instance;[69] particularly on the droves from Gisborne to the Manawatu and Gisborne to the Waikato through the winter months: they offered comfort hitherto unknown on a drove. Arthur McRae laughs at the memory of the change it wrought: 'It shouldn't have been allowed – too good for a working man.'[70] For Bill Pullen, the caravan meant that his wife and two boys could accompany him while droving: 'It was bloody good. It was the ultimate droving scene . . . it was drover's heaven.'[71] Still, Gill Pullen remembers a misjudgement they made when camping on the Awatere riverbed. A safe spot when it was dry turned out to be an island when it rained.

> Where the river is you go off the road and you drop right down on
> to the riverbed and then we pulled back up onto a nice high piece of

ground . . . it turned out to be an island. . . . we kept an eye on the
river and it didn't seem too bad and we went off to sleep. . . . And then
about midnight [Bill] sort of thought 'I better have a look.' And then
he looked out and all he could see was water . . . So he . . . opened
the door and stepped outside up to his knees in water. So he thought
'God we've got to get out of here.'. . . So he jumped in the truck and
of course took off to get to the road but, of course, we had to go down
into the river before getting up. I could hear the truck coughing and
spluttering and I was in the caravan with the kids and watching the
water coming up through the door and thinking, 'Oh my God.' But
anyway the truck gave another splutter and a cough and away she
went, she got traction and up and onto the road. [72]

The joy of the narrow escape was soon forgotten when Bill bogged the caravan in a council metal pit and the family spent an uncomfortable night on such a lean that staying in bed was a feat. A kindly cockie towed them out in the morning.

No matter the era, life on the road was rough and minimalist. Only the essentials were taken on the road, and the margin between being clothed and fed and being stuck, cold, wet and hungry was often a bare whisker. For some of the drovers in later years, living in a caravan was like being in the lap of luxury – especially when the alternative was sleeping wrapped in canvas on the hard ground – but even then it was still a demanding and rough life that required a willingness to make do with rudimentary, no-frills conditions that were in sharp contrast to our lives today.

A Recognition

I've been droving when the weather was both boisterous and cold,
And likewise when the sun was like a ball of molten gold.
In the gloomy days of winter, I've been driving on the track;
Also when the sun was blazing till my skin was like to crack.
I have felt fatigue and hunger, while I've driven weary sheep
Where people would begrudge a bit for man or beast to eat.
But I lately went a-droving, past a station I could name,
Where the river had got flooded by the recent heavy rain.
I was baulked and in a muddle, I was puzzled what to do.
I had come a weary distance, and my sheep were starving too.
I pulled up at the station in a doubtful kind of way
For I knew that they were bothered with intruders every day.
It was also crutching season, so their grass could not be flush
And their yards were very apt to be knee deep in mud and slush.
But I had to face the music and must chance it so to speak
Though I felt like an intruder who was travelling on the sneak.
But my fears were quickly banished, and I felt I need not quail,
As I heard the welcome echoes of a very friendly hail.
Someone sang out 'Dinner's ready! Come along, you're just in time.'
And just a minute afterwards a bell began to chime.
I had a hearty dinner, my sheep were grazing too,
For I left them in a paddock that the highway passes through.
Then I loitered round till evening, when the bosses hove in sight,
For I meant to humbly ask them if they'd let me stay the night.

> So I told them my situation, and, answering with a laugh,
> They told me I was welcome, amidst banter fun and chaff.
> I wasn't ordered onward, but made welcome there instead
> To the very best of eating and a warm, cosy bed.
> And grass – my sheep were given the best pasture that they had.
> It saved the life of many that were looking weak and bad.
> I was treated with a kindness I had seldom met before.
> Had I been a wealthy squatter they could have done no more.
> Twas the old Otairi Station, and my thanks I freely give
> To its owners, whom I'll think of for the longest day I live.
> May they always have the power to be generous and free
> With their kindly hospitality, just as they were to me.
>
> — Anonymous drover[73]

Drovers moving a mob of sheep on Mt Cook Road in Canterbury. *Andris Apse*

9.
THE
TWILIGHT
YEARS

The growth in trucking and the
demise of droving

It was the day of the truck.[1] Cattle, sheep, horses, pigs and goats have all been driven on the hoof along New Zealand roads at one time or another – on rural roads, state highways and even suburban streets; sometimes in a quiet, orderly manner and other times in a storm of chaos and confusion.

The drover, over the years, has seen the landscape evolve as the country developed into an agricultural and pastoral powerhouse, producing quality export goods. They have experienced first-hand the impact of pastoralism, the goldrushes, agricultural and technological advancement – from simple things such as fencing, to the development of the frozen meat trade, aerial topdressing and soil husbandry. The benefits of this enabled the Government to open up once marginal land for rehabilitation and ballot farms after World War Two, and this led to thousands of sheep and cattle being driven to newly developed blocks. Infrastructural progress enabled the drover to move more freely along their regular routes. These 'stock routes' became enshrined in county and borough councils bylaws, giving the drover right of free movement.

Such development, however, was a double-edged sword. As isolated regions were opened up with better roads, a growing trucking fleet could carry stock more easily to market, and this, coupled in the 1960s with

A far cry from the roads of the twenty-first century: those droving on this road near Raetihi in 1909 were faced with a muddy challenge. Drovers bore witness to the ever-evolving landscape and were to benefit from developing infrastructure. *Sir George Grey Special Collections, Auckland Libraries, AWNS-19091014-10-2*

changing Government policy allowing for trucking to compete with rail, meant there was less demand for droving. At the same time, the dealer or trader's role in providing the drover with a ready supply of work became less important; and with growing urbanisation, traffic increased and stock became less welcome on busier roads. Gradually all these changes impacted on the viability of the drover's work. Today it is a novelty to see stock being driven, and many of those who now move the occasional mob would know little of the long tradition of those who have gone before them.

Trucking

Trucking really got under way in the 1930s as trucks became more reliable, with bigger engines, increased tonnage, heavier suspension and pneumatic tyres. Lorries were simple flat-decks, with wooden detachable stock crates – mostly double-deck for carrying lambs and sheep. Stock-carrying capacity was limited to 100–120 lambs or 80 ewes – far fewer than on a drove – but the advantages of trucking included shorter travelling time and less stress for the animal. Cattle were also carted by truck, but the limited carrying

Drivers for Barwood Motors stand proudly in front of their trucks at the Tekapo sale, circa 1951–2. From left to right: Clutha Stanley, Les Borrell, Alec Bateman, Jack Cassie. *Barwood Family Collection*

Two methods of moving stock jostle for position – the old and the new. Moving stock 'on the hoof' was still a regular occurrence on the East Coast until the late twentieth century. Even as early as the 1940s, when this photo was taken, there was a sense of inevitability about trucking's eventual domination of the industry. *Archives New Zealand, AAQT 6401/A1860*

capacity – 8–12 beasts – meant that big mobs were still moved on the hoof for a considerable time.

In the 1950s, articulated truck and trailer units with increased carrying capacity were introduced; and the 1970s was the era of the big rigs. New Kenworths, Mercedes-Benz, Volvos and Macks could take, with trailer, around 650 fat lambs or 400 ewes; a single deck would take approximately 31–32 cattle; and a double took around 42 cattle beasts.[2] Often three-deck sheep crates were converted to two-deck cattle crates.

There were several obstacles to the successful development of the livestock trucking industry in the first decades of its existence. While it was still in its infancy in the 1930s, the Government passed legislation that restricted its ability to compete with New Zealand Rail. This effectively reduced it to local movement or moving stock from farm to railhead, except in a region where there was no rail service. World War Two affected the growth of trucking, too, with fuel restrictions, tyres in short supply, and the Government's ability to second vehicles. In addition, Government spending was diverted away from infrastructure, and inadequate roading in rural areas meant cartage was often difficult. It wasn't until the 1950s that the Government began to spend money on improvements. In the early 1950s, rail was rationalised and a number of branch lines were closed down. A combination of this and the freeing up of stock transport regulations in the early 1960s meant that the stock trucking industry was finally able to flourish.[3] Of course, these developments had an effect on the availability of work for the drover.

Some drovers felt no animosity toward the truckies; they appreciated their assistance along the drove:

> You see, you'd know them – they're going backwards and forwards through your mob . . . you'd give them a dollar, 'bring us back a loaf of bread' or a cabbage or whatever you wanted and if they were all pretty good, you know, and if they weren't coming through they'd give it to another truckie that was going up your road and they'd pass it out to you . . .[4]

Others resented their intrusion:

> When the trucks came in it took our living away from us. It wasn't nice at all. The work dried up. It wasn't worth staying on the road to do the big mobs. There might be a week go by and I didn't do anything. It didn't suit me or the dogs. They'd go lame when they went back on the road. Their feet get soft and the pads come off.[5]

Trucking companies popped up everywhere – though not all were made equal:

> If you put your cattle on an Arnolds' truck you knew it was going to get where it was meant to. Road Services were the tail-end of the Road Service complex in New Zealand and they got all the hand-me-downs . . . We sent out four trucks of lambs with Road Services one day and only one got to Ross. The other three broke down on the way.[6]

The beginning of the twilight years

It is difficult to pinpoint exactly when droving entered its twilight years. By 1927 there was a growing sense of the tide turning in favour of more modern transport: 'The motor car has almost driven the horse from the road. It would appear, too, that the days of the drover are numbered, for the transport of sheep by motor lorry is fast gaining favour . . .'[7]

However, like any new technology, the uptake of trucking varied according to individual circumstances. There were farmers who found the arrival 'a godsend', as did Erle Riley, farming up at Farewell Spit. Fred Cowin, not far away from Erle at Paturau, also appreciated the arrival of trucks:

> The joys of droving and the romanticism of it is greatly overrated – toward the finish there was nothing but worry and starvation for man and beast. People who had holding paddocks had sold them, the council had disposed of all county holding paddocks or they'd fallen in to disrepair where you couldn't use them . . .[8]

Bruce Ferguson, again not far away in Kaihoka, appreciated trucks; his wry comment was: 'I'd drive sheep if I had to.'[9] All three men lived on isolated farms and to have trucking access, even with limited capacity, felt like progress. John Sullivan on the West Coast recalls the change to trucking for their family: 'It just happened really . . . sort of overnight really, in the period between late '46 and mid '47.'[10]

Anecdotally, at least, it would seem that by the mid-1950s in most areas droving was well on its way out. Peter Cloake, working as a stock agent for Dalgety's in Feilding in 1953, remembers the drovers still working, but also that trucks were well established. Sonny Osborne of the Manawatu comments in his memoir *Droving Dogs and Sheep*: 'By 1939 and into my eleventh year of droving, I was quietly phasing out, doing 40 days only that year . . .'[11] Census figures concur with this: drover employment figures trend steadily downward from a high of 976 in 1936 to 447 in 1945 and only 187 in 1951.[12]

The growing rural–urban divide

In pre-1900 New Zealand, nearly 60 percent of the population lived rurally; by the 1940s it was less than 40 percent; and by the 1980s less than 20 percent were rurally based.[13] Owen Gibbens's Uncle Harry was a drover, and drew Owen's father into droving.

> . . . when you say no farming background the situation was that the areas were virtually rural, everything was rural, you were born and most lived on a small holding – you know, four or five acres of land and a cow and ducks and the dogs and all the other things that go with it. I think you would say in the times we were more in tune with the land, and it wasn't as today where you're a townie and you're a farmer – it wasn't like that.[14]

This shift to a fast-growing urban population with an increased disconnect from the rural lifestyle meant less tolerance toward stock on the road. Drovers recall increased impatience and abuse, with comments such as: 'Haven't you heard of trucking?' or 'Who's going to wash my car?'[15] David Stroud recalls: 'I got abused quite often simply by being on the road, for no other reason . . .'[16] Not that at times there was not some reciprocity: Keith Sowerby, droving in Feilding, had his own answer to motorists: 'If he got annoyed with motorists he used to really back up the cattle on them . . . He was there before the car came type of thing.'[17] The drover's standard response to the abuse was: 'They've got to remember these roads were built by cattle, the dairy cow and the animals; now it's tyres and diesel . . .'[18] Arthur McRae reflects that even those who live on the land now no longer have the same connection.

> It's the younger ones, you see, they don't know. I mean a lot of the young farmers now – their fathers were drovers and, you know, they did what I was doing . . . They've been to university and they got a degree and they've done their overseas travel and they can't see why they should put up with cattle outside their place walking on their road. I said to them, I said, 'Well, cattle made these roads: it wasn't even a track here till cattle started walking . . . [and] while you're discussing it with him your cattle are feeding like hell.[19]

One of the ironies in this divide – and another nail in the drover's coffin – was when city dwellers began to move out to lifestyle blocks. By the 1970s there was an increase in the number of small holdings (properties of less than 10 hectares).[20] But not all these newcomers took so readily to everything that the semi-rural lifestyle entailed. Drovers recall lifestylers buying cheap land and then protesting at having stock passing, making a mess on the road or on their frontage. Bill Pullen comments on the effects of this:

> The whole thing – there was a snowball effect. To give you an example, like, coming through Taneatua . . . the stock route was around the back of the village . . . a sort of lane . . . so you could go around there without annoying anyone or standing on any pansies or any bloody thing. And everything was kosher.
>
> Well then, of course, as the village grew the council tarsealed part of the laneway and sold the sections on the other side . . . when they got their little dwellings and that erected, it suddenly became a street and they suddenly didn't want drovers on their bloody stock route – but they were quite happy to buy the cheap sections.[21]

And Arthur McRae adds:

> What had happened, there was a stock route – you can go through Ashhurst, you came over the top of the hill round hard to the right and up and went on the back roads to Colyton. But then, there were no buildings – it was a metal road, no buildings, then some clever dickies buy land there, get a permit, a building permit, they build a flash house. Then it's too . . . dusty in the house so they want tarseal, so they tarsealed it and then when it's tarsealed they want to stop droving stock – they don't want droving stock to go past our house and walk on our lawns. You know that sort of thing that happens everywhere. If you buy land on a stock route you buy it because it is on a stock route and it is cheap and you can't build on it . . .[22]

In the drovers' experience, councils listened to the public outcry and changed bylaws to place greater restrictions on them. The old-timers' resentment is plain as they tell their stories: 'In those earlier days the cockies were the council, and that changed and now even the office boy can be the big rangatira at times and he's only just left high school.'[23] In some areas, councils required drovers to apply for a permit and pay a bond – Bill Pullen recalls having to pay a $2000 bond to go through Hawke's Bay County. Not everyone took any notice of this, however:

> In later years you had to get a permit. It started in about '65 or '70, especially in the Whangarei county. Ken Lewis had a lot of rows over that. He had to deliver cattle to Kamo and he defied them. There was hell to pay! That's when he had the big article in the *Herald*. He was interviewed about it and he said, 'Stock made the roads, now motor cars own the roads and a dog's only a fly on the road . . . People blow their horns at them but when they're hanging up in the butcher's shop they'll eat them!'[24]

Councils were also increasingly concerned about the environmental impact of cattle on the road – the pollution of waterways from cattle wandering in roadside drains; the damage to roads from stock effluent – and so they increased restrictions: '. . . it's ridiculous, absolutely ridiculous – some greenie sods that don't know anything haven't been anywhere and people listen to them cos it's the squeaky hinge that gets the oil . . .'[25]

So far as the drovers were concerned, their mobs did the council a favour by cleaning up blackberry and other weeds along the roadside. Guy Shanks was scathing of how things had changed.

> The council never bothered about you in the old days; only in later years the greedy buggers on the council went on about you damaging the road. The cattle didn't damage the road! They kept the roadside clean and trimmed the trees and kept the fern and weeds out of the drains.[26]

The response in the Waikato in 1975 to a mob of several thousand sheep being driven from the Central Plateau through to the Horotiu freezing works, however, shows how droving had become an unacceptable method of stock movement. Farmers, at a time of falling sheep prices, were finding that trucking cull ewes to the works was uneconomic, so they experimented with droving the sheep 20 days to the works. The furore over the drove is evidenced in the headlines splashed over the pages of the *Waikato Times*:

> 'Drover quits sheep trek as losses mount'
>
> 'Sheep will be diverted from main highway'
>
> 'Legal steps to control drives to be taken'
>
> 'Traffic men may halt big sheep drive'
>
> 'Farmers deny claims'
>
> '4000 sheep begin "trek" to Horotiu'

It was a regular feast for reporters.

With traffic volumes of around 6000 vehicles a day the Transport Department were concerned about the disruption to traffic flow;[27] the SPCA inspectors were demanding assurances that animal welfare would be maintained;[28] and the carrier firms were crying foul that their industry would suffer.[29] While the sheep made the trek, with diversions from state highways being found, as a result of this and other planned droves local authorities in the district were quick to look at reviewing existing bylaws allowing such droves.[30]

The decline of the trader–dealer

The role of trader–dealers was coming to an end, too, and this of course meant less work for drovers. Peter Cloake, stock and station agent for

This scene of a drover with his horse and dogs out on New Zealand's roads is now a novelty. Even in 1999, when these Wairangi Station cattle took to the road, the drover's day was numbered. This was to be the last Wairangi Station drove. *Tairawhiti Museum, 501-51-A, photo by Dudley L. Meadows*

Dalgety's in Feilding from 1953 to 1970, comments that the cost of rail is what killed the dealers along with a changing farming dynamic. Originally:

> . . . a lot of the cattle that was sold through Feilding, station cattle, was dealers' cattle. If it wasn't for the dealers bringing the cattle in, well [the cattle] just wouldn't have been there, because farmers didn't like travelling up to Gisborne to buy 20 or 30 cattle that they needed in the spring . . .[31]

By the 1950s a 'younger breed of farmer' became more open to travel and went further afield to source their own stock. Peter took farmers up to the King Country to source cattle 'they were a bit of a mixture back in the '50s but everything had a price and you could go up there and buy some quite good cattle and buy them reasonably cos you didn't have that Hawke's Bay competition.'³² As such patterns emerged the dealer faded out.³³

As well, more farmers built stock-loading facilities on their farms, and chose not to move their mobs on the road.

There was, of course, still work in and around the saleyards bringing in stock from holding paddocks, drafting and penning. David Stroud took up droving in Feilding on sale day as late as 1970, and carried on working there for seven years. Throughout this time he also picked up small droving jobs in and around the district, which helped to augment his income through tough economic times.

Droving's demise was in no way systematic – each region's droving patterns had varied – so it is little surprise that there was no uniformity in its decline. The North Island was slower to establish its strong farming roots, but once it was settled, stock numbers quickly overtook those of the South Island. This, coupled with the higher proportion of sheep to cattle in the South Island, meant the end to droving came quicker in the south. As long as it was economic and convenient, the East Coast's unique position as a high producer of store stock in the North Island meant stock, particularly cattle, were being driven out of the region much later than elsewhere in the North Island. Eventually, however, even this region succumbed to growing pressure to allow the old ways to rest.

EPILOGUE

It is going on 200 years now since the first sheep arrived – and survived[1] – on our shores. They were brought here by missionary Samuel Marsden. Since that small beginning, farming has long been considered the 'backbone' of our country's economy. Drovers were an important cog in the wheel of this industry, ensuring millions of stock were transported to their markets – as Don Monk reflects, 'it was a big responsibility. When you think of all those cattle and how much they were worth, well they were worth £100 in them days. Well that was a mighty lot of money you were responsible for.'[2]

When drovers reflect back on their droving days, there are memories of the good and the bad. Not everyone enjoyed it – 'it was the most boring bloody job just walking behind a mob of cattle all day or sheep'[3] – but many more did. It gave them the opportunity to be their own boss, to have 'nobody nagging in your ear',[4] to be out meeting people, and to have pride in their work:

> . . . and the stock agents, if you've got any reputation for the quality of your work they all sort of look up to you. Oh Jesus, 'How are you going? Got plenty of work? If you're ever short of a job give us a ring.'
> . . . And to me that's really something.[5]

Laurie McVicar, as a lad, would have paid to go if need be: 'It suited me. I enjoyed it . . . The thing is when you're dealing with horses and cattle they're not all that smart, and I was dealing with my intellectual equals.'[6]

There are still pockets of the country where a drove might be carried out, but the days of droving have all but disappeared. Gone are the queues of mobs heading in to the freezing works; or the days of drovers trying to outwit each other to get the best feed for their stock; or leaning on a farmer's fence, passing the time of day while stock amble by; or children plonked on their pony and told to ride out in front of the mob. We live in a different age now, where speed and time are of the essence.

As to those who drove, it had become an old man's game; the young guys saw no future in droving and took to other work. And as droving began to fade away:

> . . . [the old guys] fizzled out happily. The odd one, like old Smokey Thompson – he carried on till he couldn't ride a horse, silly old bugger. He was only one of a few. Trevor Stewart, he drove, I think, from when he could walk until when he couldn't stand up. Poor old bugger . . . He drove more stock out of Gisborne than most people. Even his honeymoon was on the road with his wife.[7]

Those who remember their droving days certainly do not blow their own trumpets, but there is a sense of pride in the work they did. Hugh Monahan, who assisted Joe McClunnen out of Murchison in the 1930s, comments on his pride that as a lad of 17 or so he could 'help someone senior like that'.[8] Colin Bass, a schoolteacher who was eager to experience droving, recalls that he 'just loved doing it. I loved cracking a stockwhip, riding a horse, chasing cattle and having an excuse I suppose to get on a horse.'[9] Bill Pullen recalls the love for his work, the 'new views every day' and the sense of being 'born' to drove.[10]

When you ask a drover about their droving days, a light comes over their face as they enthusiastically relive their time on the road as if it were just yesterday. You get a sense that, given half a chance, they would be saddled and ready to leave again in an instant.

THE SONG OF THE DROVER

The roads that we travel are many,
    They wind o'er the range and the plain;
My pack and my dogs and my neddy,
    We know the old places again.
Each bank on the way is familiar,
    Each creek is the same as of old;
We've been there in summer and winter,
    In dust and the rain and the cold.
Once more to the upland we're trekking
    I trundle along at the tail;
My lead at the head, where' he's checking
    The mob to the pace of a snail.
The sheep on the hot road are stringing,
    But not a green bite do they find;
The bark of the dogs wildly ringing
    Keeps stragglers from lagging behind.
They've not had a feed worth the calling
    These five sweltering days on the road;
It's a wonder there's none of them falling
    And making a tidy pack load.
The pack horse would be in commission,
    And so would my sleepy old hack,
The gear would then be in addition –
    I've known him take three on his back.
The trail o'er the mountains is dreary
    When autumn is sleeting and cold;
The lambs are all lagging and weary,
    And nibble their wool in the fold.
The wind o'er the saddle comes sweeping
    In gusts that are laden with hail,
The creeks are all foaming and leaping,
    And dance to the roar of the gale.

Such times are enough to dishearten
    The mob struggling on in the rain,
It's hard then to get them to start on
    The road leading down to the plain.
Along the old coast-road we're trailing,
    The dust in white clouds rising free;
A far-away, quaint, sullen wailing
    Comes up from the depths of the sea.
Past willows with long branches drooping,
    That swing as by light breezes fanned;
Past blooms of the gorse and the lupin
    That grow on the road o'er the sand.
A glimpse of the rich, sapphire ocean
    Is seen through the dips of the dunes,
Its ridges, white flecked, all in motion,
    Roll up to the beach with their tunes.

*

I still hear the deep throated ocean
    When crooning its old lullaby,
As rolled in my rug in the open
    I gaze at the star-studded sky.
All tranquil and still, save the bleating
    Of ewes that have not found their lambs,
And crunching of horses still eating
    The rushes that grow in the sands.
I think of the days I've been roving
    These same roads for year after year,
Of mates who have been with me droving,
    And yarns by the camp-fire's bright cheer.

—Hugo Finn, drover in Gisborne–East Coast,
early 1900s[11]

Sheep are unloaded from trucks and prepared for sale at the Levin saleyards in 1958. *Alexander Turnbull Library, PAColl-6303-02*

CONTRIBUTORS

Featured below are small biographical accounts of all those I have interviewed in the course of this rather large and long project. Between 2003 and 2010, I conducted some 60 interviews of varying lengths, some gathered while travelling on holiday with my family throughout much of the country. This is only the tip of the iceberg; there were many more I could have interviewed, and there were some regions I unfortunately did not reach; at some point you have to say enough, though I admit I find that difficult.

I am indebted to each person who took the time to share their memories with me. The oral history interviews may be found at the Oral History Centre of the Alexander Turnbull Library. The material included in this book is only a smidgen of what was shared with me, and I would encourage those who are interested in New Zealand's farming history to explore the stories further at the National Library.

NORTH ISLAND

Donald Trevor Beard: b. 1935 *Interviewed: 2004*
Born Feilding. Don grew up on the family farm on the Kimbolton–Rangiwahia road out of Feilding. He eventually took over the family farm after an 'apprenticeship' working on South Island stations. He never went droving, but he had recollections of Charlie Sundgren and droving in and around Feilding. Don was one of two people instrumental in starting my work interviewing drovers. He was chair of the committee for the 'Drover and Dog Project Incorporated', raising funds for a statue of a drover and his dog to be erected in Feilding.

Thomas Henry Lial Caldwell Bredin: b. 1927 *Interviewed: 2007*
Born New Plymouth. As a local New Plymouth historian, Lial had taken to collecting stories from around the district. He was enlightening when it came to district droving stories, and also had a few of his own. As a lad growing up on a dairy farm in Warea he moved stock to run-offs, and as a dairy farmer he drove his cull stock to the local saleyards.

Alben (Alby) Charles Burney: 1922–2012 *Interviewed: 2004*
Born Christchurch. Alby lived in Feilding from the mid-1940s. He drove in the Manawatu; Rangiwahia and Apiti were regular haunts. Alby worked in and around the Feilding saleyards droving stock to holding paddocks, yards and the railhead for nearly 40 years.

Phyllis Irene Clements: b. 1926 *Interviewed: 2008*
Born Whangarei. Phyllis's father was a farmer, drover, cattle dealer and butcher. From a young age Phyllis worked alongside him: a keen horsewoman, she helped her father after leaving school in Standard Six, droving around Matapouri, Tutukaka and Ngunguru. After she married and shifted to Kaitaia she worked full-time for Ken Lewis, droving on his spring cattle droves up until the last one in the early 1990s.

Peter William Cloake: b. 1935 *Interviewed: 2003*
Born Wellington. The family moved to Palmerston North in 1942. Began work as a junior aged 18 at Dalgety's stock and station firm, and worked his way up to a stock agent position in Feilding. Peter had great memories of working at Feilding saleyards, and much insight into the drovers working in the region. He explains the relationship between stock agents and drovers.

Allan Crawford: b. 1935 *Interviewed: 2008*
Born Te Awamutu. Allan grew up on the family dairy farm in the Waikato. The family shifted to Okaihau in 1946. He took on his own farm after leaving school. Farming was Allan's life until Rogernomics impacted in the 1980s; needing to supplement his income he took up droving and mustering. He went into competition with Ken Lewis on the Far North spring drove, driving his stock further south than Ken, to Paparoa. Allan drove for about five years until the Far North drove came to an end because of changes to how the Lands and Survey blocks were run in the region. His daughters Elisabeth and Anna helped on the droves.

Jack ('Boy') Leslie Curtis: b. 1937 *Interviewed: 2007*
Born Gisborne. As a young lad in the school holidays Jack was out with his father droving stock from the backblocks of Gisborne into Matawhero saleyards. His high-school days were split between school and shepherding. By age 16, school was left behind and droving was his focus. Jack drove around East Coast, Bay of Plenty and through to the Waikato. When there was no droving work he would go mustering. Droving eventually took a backseat to farm ownership, and by the 1960s when he was married with children his droving days were largely over.

Spencer John Alexander Dillon: b. 1929 *Interviewed: 2008*
Born Wanganui. Spencer grew up around stock; his father was a shepherd and manager of stations in the Rangitikei. As a shepherd at Duncans' Otairi Station, Spencer (aged about 18) drove stock to and from the railhead. After his marriage he moved to Hastings where he drove locally full-time for five and a half years. As droving work reduced he took to a variety of shepherding and farming jobs before getting a ballot block in Putaruru and, later, a dairy farm in Tirau.

Henry (Harry) William Frank: b. 1923 *Interviewed: 2008*
Born Tangowahine, Northland. Farming life was all Harry knew from a young age. He left school aged just 12 and worked on a variety of farms until, aged about 18, he decided to go droving. The Rehab blocks around Rotorua were being developed and stock was on the move in the hundreds. He drove around the Central Plateau, East Coast and Waikato regions. Harry's droving days came to an end in the early 1950s when he married and the 'kiddies' came along.

Albert (Bim) James Furniss: 1927–2010 *Interviewed: 2008*
Born Hamilton. Apart from a short stint in Auckland, Bim (nicknamed after a monkey in Auckland Zoo) spent his life in the Waikato, growing up in Ruawaro. At 13 he took to the roads droving and working on the family farm for his father, due to labour shortages as a consequence of World War Two. Up until the age of 22 he moved cull stock that his father, a trader, bought up around the district, droving them to the works or to the sales at Te Kauwhata or Ohinewai. As trucks came in, droving opportunities lessened and Bim took on farming full-time.

Contributors

Lee Fyers: b. 1940 *Interviewed: 2008*
Born Auckland. Lee's father, Buckley Fyers, was a dealer–drover and farmed in the Waikaretu Valley, Waikato. From 1948 until the 1960s he bought up stock on the East Coast and had them driven across to their farm. Lee himself went on short local droves, but never took on the drove from Gisborne. Lee recalls the drovers bringing stock across from the East Coast and some of their idiosyncrasies.

Malcolm Campbell Grayling: 1931–2010 *Interviewed: 2007*
Born New Plymouth. Fitzroy, New Plymouth, where Malcolm grew up was semi-rural at the time. Malcolm rode a pony to school aged around 10, and by 11 he was involved in droving. Malcolm's neighbour Mark Barnett, a local stock dealer, drew Malcolm and his older brother Phillip into droving his stock locally at weekends, after school and during the holidays. Both Malcolm and Phillip went farming after they finished school.

Mike and Andy Hurley: *Interviewed: 2003*
Mike and Andy Hurley are third-generation Rangitikei–Manawatu farmers. This interview centred on the development of the Hurley family farms and the resultant stock movement. They recalled the cattle droves from Gisborne to the Rangitikei hill-country stations Siberia and Papanui. They also had memories of the cattle drove to Feilding for the big annual 'Hurley sale'.

Sam Jefferis: b. 1951 *Interviewed: 2008*
Born Waerenga. Sam grew up on the family farm in the Waikato. He recounts how his father, Jack Jefferis, over 40-some years, bought up store stock from Northland and the East Coast and had them driven down onto the Waikato farm. It was a big production that employed a large number of drovers. Sam recalls being involved in some of the droving in the later years, and recalls a trip coming across from the East Coast.

Hazel Joyce Carol MacKenzie (née Gibbs): b. 1927 *Interviewed: 2007*
Born New Plymouth. Carol grew up on the family farm in Urenui, North Taranaki. Carol has lived and worked on the farm her whole life. Carol recalls drovers coming through their property on Clifton Road to go through the Te Horo tunnel and down along the beach at White Cliffs to avoid droving on Mt Messenger. She remembers helping on the droves as a girl; and the drovers staying at their home.

Arthur Francis Pender McRae: b. 1931 *Interviewed: 2008*
Born Wellington. The son of a policeman, Arthur took any opportunity to work on relatives' farms as a boy. After finishing school in Wairoa he started out as a cowboy, then moved to working as a shepherd and took up some droving work in the district. He recalls how his droving work developed, becoming a respected stockman working the East Coast down through to the Rangitikei–Manawatu area. Arthur was one of the last working drovers in the Gisborne area at the beginning of the twenty-first century.

Noel Harold Martin: b. 1936 *Interviewed: 2003*
Born Eltham. Noel grew up on the family dairy farm in Eltham. He left school aged 16 and worked as a shepherd, including on Duncans' Otairi Station, where part of his role was droving stock to the railhead. He drove with Linc Campbell bringing stock south from Gisborne for three seasons. He also drove stock from Wairoa to Ohakune on his own account, as a trader, for two seasons.

Neil Matthews: b. 1933 *Self-recording: 2008*
Born Kaitaia. Neil grew up on his family's dairy farm. By age 13 or 14 Neil was droving family dairy stock to and from their run-off. Neil picked up more droving in his thirties, working for his neighbour Ken Lewis, and droving on his own account for local farmers. In the 1970s when farmers were moving out of dairying, he drove a lot of dairy herds to the saleyards. He also worked at the Kaitaia saleyards on sale days, bringing stock into the yards and taking them away after the sale. Neil's droving days finished in the 1980s, when the rules and regulations became too tough.

Benjamin Joseph Morunga: 1929–2012 *Interviewed: 2008*
Born Hokianga of Te Rarawa descent. Ben recounts memories of working with his father bushfelling when young. As bush work declined, the focus was more on stock work and developing the family farm at Panguru. He recalls taking on droving work to help pay the farm mortgage. He drove for Lands and Survey farms, Dalgety's, and North Auckland Farmers. He eventually became manager of a Lands and Survey farm, and continued to drove in the Paponga–Broadwood area.

Christina Margaret Morunga (née Brocket): b. 1954 *Interviewed: 2008*
Born Paeroa. Christine's recollections of her early life include time spent on the family farm; she recalls moving stock with her father when she was 12 on their Ruapekapeka farm. She did nursing training and moved to Broadwood in the Hokianga area. She tells the story of how she met and married Ben, and recalls accompanying Ben on local droves from Mitimiti to Broadwood up until 2007.

Ian Maxwell Mullooly: b. 1924 *Interviewed: 2008*
Born Gisborne. Ian was just 15½ years old when he took to droving full-time. His father, a farmer and stock buyer, always had plenty of stock to move. Initially Ian drove alongside his father and his business partner. Eventually he took over the droving, travelling all around the East Coast and Bay of Plenty. After three years Ian turned to shepherding and left the droving days behind.

Brian Arthur Philps: b. 1928 *Interviewed: 2003*
Born Palmerston North. Aged 14, Brian used to skip school and drove stock from Foxton to Shannon. He worked at Borthwicks Freezing Works in Feilding as a shepherd, foreman and buyer, but he also took any opportunity he could to drove.

Gill Leigh Pullen (née Freeman): b. 1950 *Interviewed: 2004*
Born Gisborne. Gill and Bill married in 1971. With two small boys in tow, in 1975, Bill and Gill took to the road as a family. They drove for a number of autumns and winters, taking cattle from Gisborne through to the Waikato. Gill drove in the lead, driving a truck towing a caravan.

William (Bill) Roy Pullen: b. 1940 *Interviewed: 2004*
Born Takapau. Bill grew up on stations on the East Coast. He picked up his first drove aged 14; and after his family moved to Gisborne he took up regular droving work. The development of the Taupo basin meant a lot of stock was moved there from the East Coast. He worked for a time in the South Island mustering on stations, then moved back north. After he married, Bill drove for a number of seasons with his wife, Gill, and their young family, then continued to drove by himself for a short time before settling to farm management.

Contributors

Guy Shanks: b. 1918 *(interview conducted by Helen Martin)*
Born Auckland. Raised in Kaukapakapa, Guy recalls his experiences droving in Northland and the Waikato. He drove full-time from 1948 until 1966. Droving cattle in preference to sheep, he details how his dogs worked the stock. He recalls how droving largely came to an end with the encroachment of trucking and councils' changing attitude to stock on the road. Guy recalls his memories of Ken Lewis and some of the Far North drovers.

Mary Stevens: b. 1934 *Interviewed: 2008*
Born Ngaruawahia. Mary married Ray Stevens in 1982. In a short interview she recounts her experiences droving dairy stock for Gypsy Day alongside Ray. With no real farming experience Mary recalls her enjoyment of the droves. The longest drove she was on was seven days, sleeping in a caravan at night: 'nothing like waking up and having your trousers frozen'.

Ray Stevens: b. 1931 *Interviewed: 2008*
Born Huntly. Ray grew up on the family dairy farm in Ruawaro in the Waikato. He started droving for his father aged around 11, during the war years, mainly moving dairy stock. In 1956 he bought the farm. While farming he took on droving, mostly dairy stock onto run-offs, sometimes on the road two or three days. He also drove sharemilkers mobs on Gypsy Day to new farms around the Waikato, and on three occasions took dairy cows across to Bay of Plenty. Ray was still droving dairy mobs into the early 2000s. He drove once on the Ken Lewis drove in Northland. Ray still musters on the Glen Lyon Station in South Canterbury.

David John Stroud: b. 1941 *Interviewed: 2004*
Born Dannevirke. Family moved to the Halcombe in the Manawatu after World War Two as his father drew a Rehab farm. David bought the family farm in 1969. He drove to supplement farming income through the 1970s, around Rangiwahia, Himatangi and Kairanga for seven years. David also worked moving stock from the holding paddocks to the Feilding saleyards on Fridays, sale day.

Kathlyn (Kath) Mary Burton (née Sundgren): 1923–2010 and Beverley (Bev) Lylie Jean Sundgren: 1935–2012 *Interviewed: 2003*
Kath born Eketahuna. Bev born Palmerston North. Both women shared their memories of their father Samuel Charles (Sunshine Charlie) Sundgren's work as a drover in the Manawatu from 1925 until his death in 1964.

Ngaire Thomson (née Smith): b. 1920 *Interviewed: 2003*
Born Feilding. Grew up on a farm on Ridge Road, Feilding; farm work was her love. Throughout World War Two Ngaire worked on the family farm. She recalls assisting her father with droving stock from their farm to the Feilding saleyards.

Robin Hassard Turner: 1923–2013 *Interviewed: 2007*
Born Taumarunui. Robin grew up on a family farm in the King Country. While he was establishing the leased farm, his father also worked seasonally as a buyer for Borthwicks. Robin, aged about 10, was out mustering, droving and loading stock on to rail trucks for his father. A family shift to Tauranga when Robin was around 16 meant the end of his droving.

Stock agents: Donald Yeoman b. 1923, Bob Belgrave b. 1931, Jack Muir b. 1925, Neil Finch b. 1926, and *Country Calendar* producer Frank Torley *Interviewed: 2003*
Donald, Bob, Jack and Neil all worked as stock agents with different firms in Feilding. This was a short interview that gave insight into the relationship between stock agent and drover. Each shared memories of different drovers who had worked in the district. Frank Torley attended this interview and shared a few of his own memories of his early working career in Feilding.

Broadwood Gathering *Interviewed: 2008*
A gathering was held in 2008 in Broadwood in the Far North of a number of local drovers: some had been on the Ken Lewis drove; others just drove locally. They shared their stories over a beer. Allan Crawford, Neville Clotworthy, Richie Coulter, John Holland, Mike Holland, Ross Mattheson, Neil Matthews, Bill McCready, Ben and Christina Morunga, Moses Moses, Robert Murray, John Peita, Clare Thomson, Alec Tracey, Bobby and Rose Wells, Skip Whittaker, Robbie Whittaker, Victor Yates, Colin Manihera, Sandy White, Bob Nelley, Phyllis & Kevin Clements.

SOUTH ISLAND

Allan Robert Barber: b. 1925 *Interviewed: 2007*
Born Whakatane. Grew up on the family farm at Nugget Bay, Southland. From 1942 until 1985 he worked for Wright Stephenson Co. at various locations. Allan was insightful about stock movement in the regions he worked in, and was able to recall the tail end of droving in Central Otago. He recalls stock being transported on the *Earnslaw* steamer on Lake Wakatipu.

Eric Colin David Bass: 1923–2008 *Interviewed: 2007*
Born Kaikoura. The son of a schoolteacher, Colin recalls attending 14 different schools as his father sought promotions. Colin became a teacher himself after World War Two as a returned soldier. When teaching at Bainham he boarded with local farmer–trader–drover Charlie Stewart. In the school holidays Colin would go out droving with Charlie; he recounts some of his experiences.

Malcolm (Mickey) David Clark: 1929–2011 *Interviewed: 2009*
Born Darfield. Mickey grew up in Whataroa. After leaving school he did casual work, including droving stock between Whataroa and Ross. In 1948 through to 1951 he drove on the South Westland droves for Nolans.

Alexander Morris Cocks: b. 1926 *Interviewed: 2007*
Born Waimate. Grew up on the family farm at Bluecliffs, South Canterbury. He was drawn into droving at around 12, moving small local mobs around Holme Station and St Andrews. During the war years he took on more droving work because of the shortage of workers. He mustered up in the Mackenzie but was still involved in droving; he drove stock as far south as North Otago. In 1953 Alex married and moved to Nelson, then Omarama; by then trucking was taking over.

Anthony (Tony) Patrick Condon: 1941–2007 *Interviewed: 2006*
Born Hokitika. A third-generation Coaster, Tony recalls having his imagination fed by watching the South Westland cattle drove going past their farm in Mahitahi as a child. As a lad out of high school he

assisted on the mustering and droving of Nolan cattle along with Myrtle Cron's cattle to Paringa in the 1950s; the cattle were then loaded onto trucks and transported to Whataroa saleyards.

Frederick Robert James Cowin: 1924–2007 *Interviewed: 2007*
Born Pakawau, Golden Bay. Fred was a second-generation farmer, farming below Paturau. He had memories of developing the farm and droving his stock off his farm as far afield as Bainham. He recalls the movement of stock in the area, and the impact of trucking on the movement of stock.

Allan Andrew Peter Cron: b. 1953 *Interviewed: 2006*
Born Hokitika. Allan was a third-generation South Westland farmer, and was involved in mustering stock off the river runs and droving stock from Nolans' Cascade runs through to Haast. His memories of life in the isolated settlement highlighted the pioneering nature of the residents and those who took on the South Westland drove.

Captain Maxwell David Charles Dowell QSM: b. 1934 *Interviewed: 2006*
Born and bred a Coaster, Max became aware of the South Westland drove through his work with Ministry of Works; it captured his imagination, and he organised to assist on a drove in 1956 – a 'once in a lifetime' opportunity.

Murray Neville Dymock: 1929–2011 *Interviewed: 2007*
Born Waikuku, Canterbury. Murray grew up on the family farm at Waikuku, milking 60 cows. He left school during the war, as he was needed on the farm. Aged 14 he took on work for stock dealer Stan Wright from Woodend, starting out as a tractor driver; he moved into droving stock for Stan, working for him for some 10 years. Murray recalls droving in Canterbury, and other memories of drovers and stock movement.

Charles Peter Eggeling: b. 1943 *Interviewed: 2006*
Born Greymouth. Peter drove the iconic South Westland drove just once as a 17-year-old school-leaver, on the last drove from Okuru to Paringa in 1961. Although he was more interested in the sea than farming, he recalled the excitement of the drove.

Milcah Elizabeth (Betty) Eggeling (née Buchanan): 1920–2013 *Interviewed: 2006*
Born Ross. Betty lived in Ross, then Ikamatua. Her family moved to Okuru, South Westland, when she was aged around 11. Betty participated three times on the South Westland drove from Okuru to Whataroa alongside her husband Charlie Eggeling in the 1940s. Her accounts go into details of the day-to-day running of the drove and her experience of it from a woman's perspective.

Bruce Archibald Ferguson: 1917–2010 *Interviewed: 2007*
Born Nelson. A second-generation farmer at Kaihoka, Collingwood, Bruce recalled farming life and droving stock off his father's farm and, later, when he owned it. He had recollections of the droving he did, and of stock-moving patterns in the region. He also had stories of some of the characters droving.

Myra Isabella Fulton (née Buchanan): 1925–2009 *Interviewed: 2007*
Sister of Betty Eggeling. Myra recalled her experiences of living in Okuru and assisting on the South Westland drove alongside family on a number of occasions. She wrote *Okuru: The Place of No Return: The Story about the History of Early Settlers of South Westland in New Zealand*, which recorded the life of her family and others in Okuru.

Owen Robin Gibbens: b. 1929 *Interviewed: 2009*
Born Greymouth. Owen recounts stories of droving days on the West Coast from memories of his father Dick Gibbens's working life, which included droving. An experienced stockman, Dick drove in the Greymouth district before his marriage, and then again when he was called on to help during World War Two. Owen himself was drawn into droving during this time as a young lad. He recounts stories of droving in and around the Greymouth district until the postwar period, and shares his memories of regular stock movement and other drovers.

Barry Lance Grooby: b. 1946 *Interviewed: 2006*
Born Motueka. Barry and his father, Lance Grooby, both drove around the Motueka area. Lance drove stock in and around Motueka after World War Two through until the 1990s. He assisted the Richards brothers droving cattle over Takaka Hill, and he drove sheep off Takaka Hill to orchards where they would graze over winter. Barry assisted his father when available. He particularly enjoyed droving rodeo horses between grazing blocks.

Karl Owen Jones: 1913–2010 *Interviewed: 2006*
Karl did not drove himself, but he was born and bred in Karamea and recalled the irregular droving of cattle out of the district from his childhood through until possibly as late as 1960. This interview highlighted the difficulties created by Karamea's isolation and its developing farming economy.

Mary Kennedy (née McBride): b. 1927 and Eric Kennedy *Interviewed: 2006*
Mary's family farmed in Whataroa area. She had childhood memories of the South Westland cattle arriving for the sale; and she also recalls other small droves through to the railhead at Ross. She remembers sale day being a big day for Whataroa. Mary's husband Eric came to the Coast in the 1950s and had a few memories of the sales.

Patrick William Kennedy: b. 1929 *Interviewed: 2006*
Born Reefton. Kennedy families have been farming Totara Flat on the West Coast since 1873. Pat has recollections of stock droving in the region. He remembers D'Urville Island sheep being punted across to Havelock and driven down Wairau Valley through the Buller Gorge to Inangahua Junction, where they were then trained to Totara Flat. He also has memories of droving a small number of family cattle to the butcher at Ngahere.

Barbara Kenton: b. 1943 *Interviewed: 2007*
Born Palmerston North. Barbara's father Keith Sowerby was born in 1906 in Feilding and drove in the Manawatu from the age of 15 until the 1970s. Barbara had many memories of her father's working life. Julia Sowerby, Barbara's sister, also offered insight into her father's work via email. Droving may well have been in the blood: Keith's grandfather Henry Sowerby was believed to be the first person to take cattle overland to Timaru.

June Leslie: b. 1927 *Interviewed: 2007*
Born Timaru. June grew up on the family farm at Mawaro, Cave, South Canterbury. A keen outdoors person, she was happier working with stock than doing housework. She left school after the fifth form and worked on the family farm during the war years due to labour shortages, droving stock between family blocks as well as to the saleyards at Pleasant Point. June also recalled her father's droving days before World War One.

Loftus Lawrence (Laurie) McVicar: b. 1929 *Interviewed: 2007*
Born Greymouth. Laurie's parents owned the Totara Flat Hotel on the West Coast. He has memories of passing drovers staying at the hotel, and stock held in the hotel holding paddock. Despite not coming from farming stock, Laurie was keen on horses and stock from a young age. He has memories of being out in the lead of mobs of cattle when he was aged just six. He recounts the stock movement in the area and the Totara Flat saleyards, and talks of working stock when older.

Hugh Monahan: 1918–2011 *Interviewed: 2007*
Born Murchison. Hugh's father was a dairy farmer in the Murchison area. Hugh worked on the farm until he was 21. He recalled droving stock down to the run-off as a teenager; and regularly helping a local drover with stock over bridges in his late teens. He remembers men who farmed and did droving; and stock-movement patterns in the area. Hugh's wife, Isabel (Hunter), briefly talks of her memories of her father droving.

Don Monk: b. 1935 *Interviewed: 2009*
Born Kaiapoi. Don was an outdoor boy from the start, and recalls a happy childhood out floundering, duck and rabbit shooting, whitebaiting and eeling. He recalls how much stock was on the move around the district as he grew up, and how he was drawn into droving through working as a shepherd for trader–dealer Bill ('Parkie') Parkinson. He recounts many stories of his droving experiences, including droving stock in and out of Addington saleyards, a 'marvellous place'. Don's droving days came to an end in the early 1950s when 'Parkie' died.

Kathleen Jean Morgan (née Cook): b. 1926 *Interviewed: 2007*
Born Motueka. Kathleen grew up on the family farm in Riwaka. The Cook family stockyards were used by local drovers to hold stock in overnight, including the Richard brothers' stock. In the 1950s Jack Richards called on Kathleen to help by leading the mob of cattle down off Takaka Hill a number of times. She was also called on to drove for local stock dealer Ken Jenkins. Kathleen had memories of other local drovers.

Gilbert (Bill) William Richards: b. 1932 *Interviewed: 2007*
Born Collingwood, Golden Bay. Bill, cousin to Harry, had recollections of Ned and Jack Richards's working relationship and their droves over Takaka Hill. While Bill may have helped move the odd mob of stock from Paturau through to Westhaven Inlet or maybe as far as Ferntown, he was more involved with clearing land for his father, Gilbert Henry Richards.

Henry (Harry) Edwin Richards: b. 1926 *Interviewed: 2007*
Born Collingwood, Golden Bay. Harry's father Ned and his Uncle Jack farmed in Paturau. Trading cattle was what they were known for. Their mobs were regularly seen being driven over Takaka Hill. Harry

would sometimes assist, out in the front of the mob as a lad alongside his father. Harry had insight into the working relationship of these two brothers and the history of the establishment of their farm.

Erle Trevor Riley: 1923–2010 *Interviewed: 2007*
Born Collingwood, Golden Bay. Erle was a third-generation farmer at Collingwood. He grew up on land at Wairakei on Farewell Spit. He started work on the farm aged 14, and drove farm cattle to Takaka to the butcher aged 18–19. He recalls droving cull cattle from Collingwood to the freezing works for Ralph King.

John Stephen Sullivan: b. 1927 *Interviewed: 2006*
Born Hokitika. John's family farmed at Fox Glacier. His father and uncle opened the Fox Hotel in 1928. John had childhood memories of watching the South Westland cattle coming past. He drove family cattle up to Whataroa when aged 10–11. In 1944 he first drove cattle up to Ross; 1946 was the last drove for his family.

Alexander William Sutherland: 1913–2009 *Interviewed: 2007*
Born Murchison. Alex grew up on the family dairy farm in the Murchison area; he took over the farm in 1935. As a boy he assisted Charlie O'Brien, whom he recalled being the main drover in the area, droving local lambs for the works 30 miles (48 kilometres) up to Glenhope railhead. The drove was a three-day affair and Alex always rode his bicycle out in the lead of the mob.

William (Bill) John Traves: b. 1921 *Interviewed: 2007*
Born Christchurch. Bill grew up on a small farm at the Levels. His father supported the family by droving full-time. Bill remembers his father's droving days in Canterbury; he recalls his father being away for three months once, droving from Tekapo to Blenheim. Bill himself was roped in to moving sheep down to the railhead at Pleasant Point and other local droves. He was far more interested in building than droving, so never pursued stock work.

White Sisters: Jessie Mary Bradley 1914–2012, Gertrude Ellen Oxnam 1918–2013 and Doris Maude Alice Nalder b. 1921 *Interviewed: 2007*
All three sisters were born at their home in the Buller Gorge. Local Murchison identities, all were involved in running the Murchison Museum. They have a wide knowledge of the district and shared childhood memories of drovers and droves in the Murchison area.

ENDNOTES

Introduction
1. Joyce West, 'The Drover', in A. E. Woodhouse (ed.), *New Zealand Farm and Station Verse 1850–1950*, Whitcombe & Tombs Ltd, Christchurch, 1950, p. 136.

1. The age of the 'Golden Fleece'
1. Peter Bromley Maling (ed.) *The Torlesse Papers: The Journals and Letters of Charles Obins Torlesse concerning the foundation of the Canterbury Settlement in New Zealand 1848–1851*, 2nd edn, Caxton Press, Christchurch, 2003, p. 143.
2. William Pember Reeves, *The Long White Cloud: Ao Tea Roa*, Golden Press, Auckland, 1973, p. 182.
3. John McGregor, 'Reminiscences: A short sketch of my life', 1863, Canterbury Museum, ARC 1900.337.
4. Ibid.
5. *New Zealand Gazette* and *Wellington Spectator*, vol. IV, issue 282, 20 September 1843, p. 2.
6. A. G. Bagnall, *Wairarapa: A Historical Excursion*, Masterton Trust Lands Trust, Masterton, 1976, p. 51.
7. R. D. Hill, 'Pastoralism in the Wairarapa, 1844–53', in R. F. Watters (ed.), *Land and Society in New Zealand: Essays in Historical Geography*, Reed, Wellington, 1965, p. 32.
8. Lovat, Lady Alice, *The Life of Sir Frederick Weld, G.C.M.G.: A Pioneer of Empire*, John Murray, London, 1914, pp. 21–22.
9. Bidwill, William Edward & A. E. Woodhouse (eds), *Bidwill of Pihautea: The life of Charles Robert Bidwill: an account of the early days at Pihautea, New Zealand and a short history of the Bidwill family*, Coulls Somerville Wilkie Ltd, Christchurch, 1927, p. 54.
10. Lovat, p. 22.
11. Bidwill & Woodhouse, pp. 54–55. Until an earthquake in 1848 the Mouka Mouka rocks were to cause problems for those travelling the coastal route into the Wairarapa. The route was well used for some 10 years and remained the preferred route for travellers and stockmen even after the Rimutaka road was open in 1856. Bagnall, p. 62.
12. Bidwill & Woodhouse, p. 55.
13. Ibid.
14. Lovat, pp. 22–23.
15. Bidwill & Woodhouse, pp. 55–56.
16. *Statistics of New Zealand for the Crown Colony. Period: 1840–1852*, Dept of Economics, Auckland University College, Auckland, 1954.
17. The Hawke's Bay separated from Wellington Province in 1858.
18. Returns from Land Office, 27 March 1855: Papers of Provincial Council, Papers laid on table, 1853–58, no. 408, cited in W. J. Gardner (ed.), *A History of Canterbury*, vol. II, Whitcombe & Tombs Ltd, Christchurch, 1971, p. 32.
19. *Lyttelton Times*, vol. II, issue 54, 17 January 1852, p. 1.
20. A. D. McIntosh, *Marlborough: A Provincial History*, 2nd edn, Capper Press, Christchurch, 1977, pp. 124–25; and Jeanine Graham, *Frederick Weld*, Auckland University Press/Oxford University Press, Auckland, 1983, pp. 41–42.
21. Most likely E. D. Sweet, Hillersden Cattle Run.
22. Edward Jollie, '1825–1894 Reminiscences', ATL, qms-1072, p. 31.
23. Ibid., pp. 31–32.
24. Ibid., pp. 32–33.
25. Ibid., p. 32.
26. Ibid., pp. 33–34.
27. Ibid., p. 34.
28. Ibid., pp. 34–35.
29. Graham, p. 42.
30. A. E. Woodhouse, *George Rhodes of the Levels and His Brothers, Early Settlers of New Zealand: Particularly the Story of the Founding of the Levels, the First Sheep Station in South Canterbury*, Whitcombe & Tombs Ltd, Auckland, 1937, pp. 112–113.
31. Alfred H. Duncan, *The Wakatipians or Early Days in New Zealand*, Lakes District Centennial Museum Inc., Arrowtown, 1969 edn, p. 6.
32. Ibid., p. 8.
33. Ibid., p. 6.
34. In his account Duncan does not specifically mention dogs, although he earlier writes of Gilbert and his collies, so we can assume dogs worked the drove as well.
35. Duncan, p. 10.
36. The Molyneux River was later named the Clutha River.
37. Duncan, pp. 10–11.
38. Ibid., pp. 10–11.
39. Ibid., p. 12.
40. Ibid., p. 14.
41. Grahame Sydney, *Promised Land: From Dunedin to the Dunstan Goldfields*, Penguin, Auckland, 2009, p. 24.
42. Duncan, pp. 13–14.
43. Laurence J. Kennaway, *Crusts: A Settler's Fare Due South*, reprint, Caper Press, Christchurch, 1970, pp. 44–45.
44. Ibid., pp. 46–47.

45. Samuel Butler, *A First Year in Canterbury Settlement* (ed. A. C. Brassington & P. B. Maling), Blackwood & Janet Paul, Hamilton, 1964, pp. 101–02.
46. Butler, p. 101.
47. W. David McIntyre (ed.) *The Journal of Henry Sewell*, vol. 1, Whitcoulls, Christchurch, 1980, p. 210.
48. Woodhouse, 1937, p. 55.
49. Kennaway, pp. 55–56.
50. Ibid., p. 56.
51. Owen Gibbens interview.
52. 'The eradication of scab in New Zealand', *Otago Witness*, issue 2029, 12 January 1893, p. 7.
53. Kennaway, pp. 130–32.
54. J. B. Condliffe, *New Zealand in the Making: A Study of Economic and Social Development*, George Allen & Unwin, London, 1959, p. 143.

2. On the Move

1. J. B. Condliffe, *New Zealand in the Making: A Study of Economic and Social Development*, George Allen & Unwin, London, 1959, p. 129.
2. The term 'golden fleece' was used in Charles Hursthouse, *The New Zealand of Today*, London, 1867, p. 16, cited in R. P. Hargreaves, 'Speed the plough: an historical geography of New Zealand farming before the introduction of refrigeration', PhD thesis, University of Otago, 1966, p. 240.
3. G. T. Alley & D. O. W. Hall, *The Farmer in New Zealand*, Department of Internal Affairs, Wellington, 1941, p. 90.
4. A. E. Woodhouse, *George Rhodes of the Levels and His Brothers: Early Settlers of New Zealand*, Whitcombe & Tombs Ltd, Auckland, 1937, p. 201.
5. Tom Brooking, 'Economic Transformation', in G. W. Rice (ed.), *The Oxford History of New Zealand*, 2nd edn, Oxford University Press, Auckland, 1992, p. 233.
6. Rae L. Moore, 'The history of Cheviot Hills 1846–1893', MA (Hons) thesis, University of New Zealand, 1937, p. 91.
7. Bronwyn Dalley & Gavin McLean (eds), *Frontier of Dreams: The Story of New Zealand*, Hodder Moa Beckett, Auckland, 2005, p. 197.
8. 'Stratification of the Industry', in A. H. McLintock (ed.), *An Encyclopaedia of New Zealand*, 1966; Te Ara – the Encyclopedia of New Zealand, updated 23-Apr-09, www.TeAra.govt.nz/en/1966/lamb-and-mutton-production/3
9. 'The Role of Beef Cattle in New Zealand', in A. H. McLintock (ed.), *An Encyclopaedia of New Zealand*, 1966; Te Ara – the Encyclopedia of New Zealand, updated 22-Apr-09, www.TeAra.govt.nz/en/1966/beef-cattle-and-beef-production/5
10. John McCrystal, *100 Years of Motoring in New Zealand*, Hodder Moa Beckett, Auckland, 2003, p. 74.
11. Ibid., p. 77.
12. Robin Bromby, *Rails That Built a Nation: An Encyclopedia of New Zealand Railways*, Grantham House, Wellington, 2003, pp. 13–15 gives a very good summary of railway development.
13. 'The development of New Zealand's railway system 1863–1963', reprinted from New Zealand Official Yearbook 1963, p. 7, viewed at www.stats.govt.nz
14. Ibid., p.11.
15. James Watson, *LINKS: A History of Transport and New Zealand Society*, Ministry of Transport Te Manatu Waka, 1996, p. 174.
16. Don Monk interview.
17. Ibid.
18. Lee Fyers interview.
19. New Zealand Official Yearbook 1901, viewed at www.stats.govt.nz.
20. Ray Stevens interview.
21. Ibid.
22. In general terms the 'season' started around August and peaked in January.
23. New Zealand Official Yearbook 1913, viewed at www.stats.govt.nz.
24. *Bateman New Zealand Historical Atlas Ko Papatuanuku e Takoto Nei*, Department of Internal Affairs, Wellington, 1997, plate 60.
25. Sheridan Gundry, *Making a Killing: A History of the Gisborne–East Coast Freezing Works Industry*, Tairawhiti Museum, Gisborne, 2004, pp. 19–20.
26. Stock agents interview.
27. Brian Philps interview.
28. Bev Sundgren & Kath Burton interview.
29. Stock agents interview.
30. Jack Curtis interview.
31. Ibid.
32. Peter Cloake interview.
33. Ibid.
34. Ibid.
35. Allan Barber interview.
36. Laurie McVicar interview.
37. George McLeod, *My Droving Days*, Ezyprint Solutions (self-published), Dunedin, 2011, pp. 2–4.
38. Peter Cloake interview.
39. Don Monk interview.
40. The Lees Valley property and Sinai were owned by Bill Parkinson and Fred Stokes in partnership.
41. Don Monk interview and telephone conversation, 2 April 2013.
42. Murray Dymock interview.
43. Laurie McVicar interview.
44. Sam Jefferis interview.
45. Bill Pullen interview.
46. Sam Jefferis interview.
47. Ibid.

48. Janet Holm, *Caught Mapping: The Life and Times of New Zealand's Early Surveyors*, Hazard Press, Christchurch, 2005, pp. 245–46.
49. Edgar Jones, *Autobiography of an Early Settler in New Zealand*, Coulls Somerville Wilkie Ltd, Wellington, 1933, p. 48.
50. Arthur McRae interview.
51. Peter Cloake interview.
52. Andy & Mike Hurley interview.
53. In the South Island, Te Anau was also being developed. Dave Ffiske, *Lands & Survey, 1950 to 1965: Northland Scrub to Grass*, D. Ffiske, Queenstown, 2008. Maori Affairs blocks were being developed at around the same time, and much stock was moved to these blocks as well.
54. The discovery of trace element deficiencies in soil enabled land redevelopment. In the 1930s, land that was unable to successfully carry stock because of 'bush sickness' – in particular, large areas of the central volcanic plateau in the North Island – were found to be cobalt-deficient. With the application of cobalt, animal heath improved and the land was gradually put under production.
55. Ffiske, p. 5.
56. Gundry, p. 124.
57. Ian Mullooly interview.
58. Gundry, p. 125.
59. Ben Morunga interview.

3. The Long and the Short of It

1. Bruce Ferguson interview.
2. Karl Jones interview.
3. Ibid.
4. Brian Hunter & Rosalie Hunter, *A Collection of Stories Representing 125 Years of the Feilding Saleyards 1880–2005: The Smell of Success*, Feilding Information Centre, Feilding, 2005, p. 5.
5. Barrie Gordon, *Matawhero, Elsewhere and Other Things: A Livestock Auctioneer in New Zealand*, B. Gordon, Gisborne, 1990, p. 31.
6. Sam Jefferis interview.
7. Gordon, p. 31.
8. Arthur McRae interview.
9. For example, Kaiapoi, Hokitika, Whanganui and many little ports up the Taranaki coast had river ports; they timed their crossing according to tides.
10. W. H. Heays, Diary/transcribed, ca1923-1924, MS-Papers-9009-19, p. 14.
11. Allan Barber interview.
12. Johnson's Barge Service Ltd email.
13. Pat Kennedy interview.
14. Owen Gibbens interview.
15. W. D. (Bill) Nolan, *The Droving Days: A History of Cattle Raising and Droving in South Westland*, W. D. Nolan, Christchurch, 1998, pp. 3–4.
16. Mark Pickering, *The Southern Journey*, Mark Pickering, Christchurch, 1997, p. 72.
17. Tony Condon interview.
18. Ibid.
19. Betty Eggeling interview.
20. Tony Condon interview.
21. Allan Cron interview.
22. Betty Eggeling interview.
23. Allan Cron interview.
24. Betty Eggeling interview.
25. Tony Condon interview.
26. Nolan, p. 35.
27. Des Nolan in Julia Bradshaw (ed.), *The Far Downers: The People and History of Haast and Jackson Bay*, Otago University Press, Dunedin, 2001, p. 79.
28. Myra Fulton interview.
29. Mickey Clark interview.
30. Tony Condon interview.
31. Nolan, 1998, p. 17.
32. Myra Fulton, *Okuru: The place of no return: The story about the history of early settlers of South Westland in New Zealand*, Myra Fulton, Takaka, 2004, p. 234.
33. Betty Eggeling interview.
34. Eric, in Mary & Eric Kennedy interview.
35. Betty Eggeling interview.
36. Mary, in Mary & Eric Kennedy interview; Mickey Clark interview.
37. Mary, in Mary & Eric Kennedy interview.
38. Max Dowell interview.
39. Mary, in Mary & Eric Kennedy interview.
40. Betty Eggeling interview.
41. *Doubtless Bay Times*, 21 May 2008.
42. Alec Tracey, telephone conversation, 24 November 2013.
43. Vernon Wright & Bruce Foster, *Stockman Country: A New Zealand Mustering Adventure*, Listener, Wellington, 1983, p. 45.
44. Phyllis Clements interview; Alec Tracey telephone conversation.
45. Phyllis Clements interview.
46. Kerry Coulter, Broadwood gathering, January 2008; Wright & Foster, p. 34.
47. 'All's Fair in the Far North', extract from 1980 summer edition of *Dalgety Digest* (staff magazine), pp. 1–2.
48. 'Last Far North droving trek seems just around the bend', Ross Barrett, *New Zealand Herald*, 22 September 1979, p. 12.
49. Wright & Foster, p. 34.
50. Neville Clotworthy, Broadwood gathering, January 2008.
51. Ibid.
52. Ibid.
53. John Holland, Broadwood gathering, January 2008.
54. Ibid.

55. Bill McCready, Broadwood gathering, January 2008.
56. Harry Richards, 'Richards Family History', 2013.
57. Harry Richards interview.
58. Richards, 2013.
59. Ibid.
60. Ibid.; and Harry Richards interview.
61. Richards, 2013.
62. Barry Grooby interview.
63. Kath Morgan interview.
64. Ibid.
65. Ibid.
66. Ibid.
67. Harry Richards interview.
68. Ibid.
69. Bill Richards interview.

4. Faces of Droving

1. Bill, in Bill & Gill Pullen interview.
2. Murray Dymock interview.
3. Arthur McRae interview.
4. Jack Curtis interview.
5. Jack Curtis; Bill & Gill Pullen; Arthur McRae interviews.
6. Arthur McRae interview.
7. Don Monk interview.
8. Peter Cloake interview.
9. Spencer Dillon interview.
10. Laurie McVicar interview.
11. Ian Mullooly interview.
12. Bim Furniss interview.
13. Robert Murray, Broadwood gathering, January 2008.
14. Ibid.
15. Bill, in Bill & Gill Pullen interview.
16. Ibid.
17. Sonny Osborne, *Droving Dogs and Sheep*, S. Osborne, Feilding, 1987, p.10.
18. Alfred H. Duncan, *The Wakatipians or Early Days in New Zealand*, Lakes District Centennial Museum Inc, Arrowtown, 1888, reprinted 1969, pp. 11–12.
19. John E. Martin, *The Forgotten Worker: The Rural Wage Earner in Nineteenth-century New Zealand*, Allen & Unwin, Wellington, 1990, p. 1.
20. Margaret Wigley, 'Ready Money': *The Life of William Robinson of Hill River, South Australia and Cheviot Hills, North Canterbury*, Canterbury University Press, Christchurch, 2006, pp. 156–57.
21. Julia Millen, *Colonial Tears and Sweat: The Working Class in Nineteenth-century New Zealand*, Reed, Wellington, 1984, p. 41.
22. Sir John Hall KCMG, 'Sheep-driving in the early days', in *Canterbury Old and New 1850–1900: A Souvenir of the Jubilee*, Whitcombe & Tombs Ltd, Christchurch, c1900, p. 126.
23. Laurie McVicar interview.
24. John Button, *Lyalldale: A Vision Realised*, Lyalldale Historical Group, Timaru, 2000, pp. 214, 262.
25. John Sullivan interview.
26. Ibid.
27. Laurie McVicar interview.
28. Ibid.
29. Brian Philps interview.
30. Malcolm Grayling interview.
31. Barbara Kenton interview.
32. Neil Matthews, self-recording, 2008.
33. By 1857, Maori in some regions in the North Island were proficient mixed farmers. Early settlers were indebted to them for their survival, as Maori provided food while Europeans established their own land. Maori agricultural dominance was eroded with the New Zealand Wars and land confiscation, along with European population growth.
34. Herries Beattie, *The Southern Runs*, Gore District Historical Society, Gore, 1979, p. 58.
35. *New Zealand Worker*, 11 December 1929, cited in Martin, *The Forgotten Worker*, p. 39.
36. William Vance, *High Endeavour: Story of the Mackenzie Country*, Reed, Wellington, 1980, p. 27.
37. Kathleen Morgan interview.
38. Christina, in Ben & Christina Morunga interview.
39. Mary Stevens interview.
40. Ibid.
41. Ray Stevens interview.
42. Bardsley, Dianne, *The Land Girls: In a Man's World, 1939–1946*, Otago University Press Dunedin, 2000, p. 111.
43. June Leslie interview.
44. Ibid.
45. Betty Eggeling interview.
46. Gill, in Bill & Gill Pullen interview; Gill Pullen, *On the Road Again: My Story of Life on the Road Droving with a Young Family*, Gill Pullen, Hunterville, 1996.
47. See Erik Olssen & Maureen Hickey, *Class and Occupation: The New Zealand Reality*, Otago University Press, Dunedin, 2005.
48. Herries Beattie, *Early Runholding in Otago*, Otago Daily Times and Witness Newspapers, Dunedin, 1947, p. 12.
49. All the material from this section is taken from oral histories conducted by the author.
50. Sundgren Family Papers.
51. Kath Burton & Bev Sundgren interview.
52. Kath, in Kath Burton & Bev Sundgren interview.
53. Bev, in Kath Burton & Bev Sundgren interview..
54. Ibid.
55. Bill Traves interview.
56. Ibid.
57. Arthur McRae interview.
58. Ibid.
59. Ben, in Ben & Christina Morunga interview.
60. Maori land had started to be actively developed for

farming in the late 1920s under the guidance of Apirana Ngata. This process of land development continued in the 1950s through until the 1990s under the Department of Maori Affairs.
61. Ben, in Ben & Christina Morunga interview.
62. Ibid.
63. Jack Curtis interview.
64. Ibid.
65. John Sullivan interview.
66. *New Zealand Woman's Weekly*, 'She runs her own farm and lonely airport', Garth Gilmore, 17 June 1963, p. 18.
67. Mary, in Mary & Eric Kennedy interview.
68. Allan Barber interview.
69. Irvine Roxburgh, *Jacksons Bay: A Centennial History*, A. H. & A. W. Reed, Wellington, 1976, p. 130.
70. Allan Cron interview.
71. Cron cattle were known for their horns.
72. Tony Condon interview.
73. Myra Fulton (née Buchanan) interview.
74. Phyllis Clements interview.
75. Phyllis Clements, Broadwood gathering, January 2008.
76. Ray Stevens interview.
77. Phyllis Clements interview.
78. Colin Bass interview.
79. Bev Sundgren & Kath Burton interview.
80. Jack Curtis interview.

5. The Craft of Droving
1. Owen Gibbens interview.
2. Bill, in Bill & Gill Pullen interview.
3. Noel Martin interview.
4. William Main letters, 2007.
5. Arthur McRae, telephone conversation, 28 July 2011.
6. Alex Cocks interview.
7. Ian Mullooly interview.
8. Bill, in Bill & Gill Pullen interview.
9. Arthur McRae interview.
10. Ibid.
11. Sam Jefferis interview.
12. Spencer Dillon interview.
13. Ray Stevens interview.
14. Ian Mullooly interview.
15. Jack Curtis interview.
16. Ian Mullooly interview.
17. Edgar Jones, *Autobiography of an Early Settler in New Zealand*, Coulls Somerville Wilkie Ltd, Wellington, 1933, p. 49.
18. Laurie McVicar interview.
19. Ibid.
20. Spencer Dillon interview.
21. Malcolm Grayling interview.
22. Guy Shanks interview by Helen Martin.
23. Robin Turner interview.
24. Jack Curtis interview.
25. Lial Bredin interview.
26. Bill, in Bill & Gill Pullen interview.
27. Sam Jefferis interview.
28. Ian Mullooly interview.
29. Sam Jefferis interview.
30. Ibid.
31. Ibid.
32. Laurie McVicar interview.
33. Ray Stevens commenting in Sam Jefferis interview.
34. Phyllis Clements interview.
35. Don Monk interview.
36. Bill & Gill Pullen interview.
37. Arthur McRae interview.
38. Jack Curtis interview.
39. Ibid.
40. Ibid.
41. Bill, in Bill & Gill Pullen interview.
42. Noel Martin interview.
43. Murray Dymock interview.
44. Laurie McVicar interview.
45. Arthur McRae interview.
46. Karl Jones interview.
47. Malcolm Grayling interview.
48. Gill, in Bill & Gill Pullen interview.
49. Betty Eggeling interview.
50. Carol MacKenzie interview.
51. Jim Rea, *Honey and Cheese*, J. Rea, Westland District, 1991, pp. 30–31.
52. *Otago Witness*, 5 July 1894, p. 22.
53. Bill, in Bill & Gill Pullen interview.
54. Laurie McVicar interview.
55. Arthur McRae interview.
56. David Stroud interview.
57. Ian Mullooly interview.
58. Jack Curtis interview; Bill, in Bill & Gill Pullen interview.
59. Bill, in Bill & Gill Pullen interview.
60. Ibid.
61. Arthur McRae interview.
62. Allan Barber interview.
63. Jack Curtis interview.
64. Bruce Ferguson interview.
65. Ibid.
66. Mo Moses, Broadwood gathering, January 2008.
67. Lee Fyers interview.
68. John Sullivan interview.
69. Bill, in Bill & Gill Pullen interview.
70. Ibid.
71. The 'club' or Working Men's Club was known as 'the office'.
72. Don Monk interview.
73. Harry Frank interview.
74. A T-wagon was a 30-foot bogie designed to carry cattle.
75. Brian Philps interview.
76. Neil Matthews, self-recording, 2008.

77. Gill, in Bill & Gill Pullen interview.
78. Sam Jefferis interview.
79. Gill, in Bill & Gill Pullen interview.
80. Bill, in Bill & Gill Pullen interview.
81. Arthur McRae interview.
82. Ibid.

6. Trials A-plenty

1. Sam Jefferis interview.
2. Herries Beattie, *Early Runholding in Otago*, Otago Daily Times and Witness Newspapers, Dunedin, 1947, p. 25.
3. Sir John Hall, 'Sheep-driving in the early days', in *Canterbury Old and New 1850–1900: A Souvenir of the Jubilee*, Whitcombe & Tombs Ltd, c1900, p. 124.
4. Beattie, 1947, pp. 25–26.
5. Herries Beattie, *Pioneer Recollections: Chiefly of the Mataura Valley*, Gore Publishing Co., Gore, 1911, p. 22.
6. Irvine Roxburgh, *Wanaka Story: A History of the Wanaka, Hawea, Tarras, and Surrounding Districts*, Otago Centennial Historical Committee (Whitcombe & Tombs Ltd), Dunedin, 1957, p. 31.
7. A. E. Woodhouse, *George Rhodes of the Levels and His Brothers, Early Settlers of New Zealand*, Whitcombe & Tombs Ltd, Auckland, 1937, p. 95.
8. Arthur McRae interview.
9. Ibid.
10. Pat Kennedy interview.
11. Fred Cowin interview.
12. Bill Richards interview.
13. Jack Curtis interview.
14. Bim Furniss interview.
15. Notes from L. Nelley, Northland.
16. Don Monk interview.
17. Ray Stevens commenting in Sam Jefferis interview.
18. Laurie McVicar interview.
19. Erle Riley interview.
20. Max Dowell interview.
21. *Feilding Star*, vol. VIII, issue 2117, 9 July 1913, p. 2.
22. *Feilding Star*, vol. VIII, issue 2121, 14 July 1913, p. 1.
23. Bill, in Bill & Gill Pullen interview.
24. Ibid.
25. Alex Cocks interview.
26. Murray Dymock interview.
27. Gill, in Bill & Gill Pullen interview.
28. Phyllis Clements interview.
29. Bill, in Bill & Gill Pullen interview.
30. Noel Martin interview.
31. Sam Jefferis interview.
32. Laurie McVicar interview.
33. Barry Grooby interview.
34. Murray Dymock interview.
35. Owen Gibbens interview.
36. Neil Matthews self-recording.
37. Kathleen Morgan interview.
38. *Feilding Star*, vol. XI, issue 2432, 27 August 1914, p. 4.
39. *Hawera & Normanby Star*, vol. XLVIII, 4 November 1924, p. 7.
40. Murray Dymock interview.
41. Allan Crawford, Broadwood gathering, January 2008.
42. Murray Dymock interview.
43. *Wanganui Herald*, vol. XXXXIV, issue 12691, 10 February 1909, p. 4.
44. Owen Gibbens interview.
45. Sam Jefferis interview.
46. Noel Martin interview.
47. Laurie McVicar interview.
48. Ibid.
49. John Sullivan interview.
50. Letters of Francis William Hamilton, MS-Papers-4328.
51. Samuel Butler, *A First Year in Canterbury Settlement* (ed. A. C. Brassington & P. B. Maling), Blackwood & Janet Paul, Hamilton, 1964, pp. 82–83.
52. John Sullivan interview.
53. Sam Jefferis interview.
54. Alex Cocks interview.
55. Murray Dymock interview.
56. Phyllis Clements interview.
57. Jack Curtis interview.
58. Arthur McRae interview.
59. Don Monk interview.
60. Mike Holland, Broadwood gathering, January 2008.

7. Partners on the Journey

1. Don Monk interview.
2. Barry Grooby interview.
3. Fred Cowin interview.
4. John Holland, Broadwood gathering, January 2008.
5. Laurie McVicar interview.
6. Ibid.
7. Hugh Monahan interview.
8. Arthur McRae, telephone conversation, 28 July 2011.
9. Carolyn Mincham, *The Horse in New Zealand: Attitude & Heart*, David Bateman, Auckland, 2011, pp. 21–22.
10. Herries Beattie, *Early Runholding in Otago*, Otago Daily Times and Witness Newspapers, Dunedin, 1947, p. 35.
11. Lee Fyers interview.
12. Jack Curtis, telephone conversation, 28 July 2011.
13. Arthur McRae, telephone conversation, 28 July 2011.
14. Don Monk interview.
15. Murray Dymock interview.
16. Ibid.

17. Ibid.
18. Rob Murray, Broadwood gathering, January 2008.
19. Laurie McVicar interview.
20. W. H. Heays, diary, transcribed, ca1923–1924, MS-Papers-9009-19, p. 16.
21. Owen Gibbens interview.
22. John Holland, Broadwood gathering, January 2008.
23. Don Beard interview.
24. Beattie, 1947, p. 22.
25. Cited in Julia Bradshaw, *The Land of Doing Without: Davey Gunn of the Hollyford*, Canterbury University Press, Christchurch, 2007, p. 85.
26. Murray Dymock interview.
27. Jack Curtis, telephone conversation, 28 July 2011.
28. Harry Frank interview.
29. Arthur McRae interview.
30. Jack Curtis, telephone conversation, 28 July 2011.
31. Bill, in Bill & Gill Pullen interview.
32. Arthur McRae interview.
33. Ibid.
34. Alec Sutherland interview.
35. Barbara Kenton interview.
36. Laurie McVicar interview.
37. Ibid.
38. John Sullivan interview.
39. Laurie McVicar interview.
40. Ian Mullooly interview.
41. Bruce Ferguson interview.
42. Harry Frank interview
43. Fred Cowin interview.
44. Lee Fyer interview.
45. Ibid.
46. Arthur McRae interview.
47. Robin Turner interview.
48. Kathleen Hawkins, 'The Drover's Dog' is an extract from a longer poem 'The Old Stock Route', 1942, in A. E. Woodhouse (ed.), *New Zealand Farm and Station Verse 1850–1950*, Whitcombe & Tombs, Christchurch, 1950, p. 144.
49. *The North New Zealand Settler*, May 1884, p. 145.
50. Skip Whittaker, Broadwood gathering, January 2008.
51. Clive Dalton, *Farm Dogs: Training & Welfare*, NZ Rural Press Ltd, 1996, p. 4.
52. Bill, in Bill & Gill Pullen interview.
53. Dalton, p. 4.
54. Noel Martin interview.
55. Arthur McRae interview.
56. Noel Martin interview.
57. Don Monk interview.
58. Bill, in Bill & Gill Pullen interview.
59. Ibid.
60. Murray Dymock interview.
61. Bill, in Bill & Gill Pullen interview.
62. Ibid.
63. Don Monk interview.
64. Bruce Ferguson interview.
65. Spencer Dillon interview.
66. Cited in Bradshaw, 2007, p. 89.
67. Lee Fyers interview.
68. Jack Curtis interview.
69. Bruce Ferguson interview.
70. Jack Curtis interview.
71. Guy Shanks interview by Helen Martin.
72. Kath, in Kath Burton & Bev Sundgren interview.
73. Owen Gibbens interview.
74. Bill Traves interview.
75. Eric Carmen, *The Broadwood Story 1888–1980*, 2nd edn, News Ltd, Kaikohe, 1983, p. 63.
76. Hugh Monahan interview.
77. Jack Curtis interview.
78. Vernon Wright & Bruce Foster, *A New Zealand Mustering Adventure: Stockman Country*, Listener, Wellington, 1983, p. 92.

8. Tools of the Trade

1. Laurence J. Kennaway, *Crusts: A Settler's Fare Due South*, reprint, Capper Press, Christchurch, 1970, p. 46.
2. Owen Gibbens interview.
3. Acland Family Papers, booklet of a 'Lecture Given by the Late G.C. TRIPP of Orari Gorge, Canterbury at Silverton, Devon 1862', Ferguson & Osborn Ltd, Printers, Macmillan Brown Library, MB44B2/41, Box 7.
4. William Edward Bidwill & A. E. Woodhouse (eds), *Bidwill of Pihautea: The life of Charles Robert Bidwill: an account of the early days at Pihautea, New Zealand and a short history of the Bidwill family*, Coulls Somerville Wilkie Ltd, Christchurch, 1927, p. 53.
5. Laurie McVicar interview.
6. Betty Eggeling interview.
7. Lee Fyers interview.
8. Murray Moorhead, *Settler Tales of Old New Plymouth*, Zenith Publishing, New Plymouth, 2005, pp. 98-100.
9. Tony Condon interview.
10. Arthur McRae interview.
11. Jack Curtis interview.
12. Bill, in Bill & Gill Pullen interview.
13. Arthur McRae interview.
14. E. C. Richards (ed.), *Diary of E. R. Chudleigh 1862–1921, Chatham Islands*, Cadsonbury Publications, Christchurch, 2003, p. 34.
15. Lovat, Lady Alice, *The Life of Sir Frederick Weld, G.C.M.G.: A Pioneer of Empire*, John Murray, London, 1914, p. 23.
16. Richards, E. C., p. 42.
17. John McGregor, 'Reminiscences. A short sketch of my life', 1863, Canterbury Museum, ARC 1900.337, p. 2.
18. Guy Shanks interview by Helen Martin.
19. Arthur McRae interview.

20. Guy Shanks interview by Helen Martin.
21. Arthur McRae interview.
22. Ibid.
23. Jack Curtis interview.
24. Bill, in Bill & Gill Pullen interview.
25. Ian Mullooly interview.
26. Ray Stevens interview.
27. Acland Family Papers, p. 4.
28. Noel Martin interview.
29. Bill, in Bill & Gill Pullen interview.
30. Hugh Monahan interview.
31. Murray Dymock interview.
32. Ian Mullooly interview.
33. Robin Turner interview.
34. Murray Dymock interview.
35. Carol MacKenzie interview: Carol remembered, with a degree of disgust, that Jimmy would 'take this soggy mess out and eat with great gusto. It looked awful but he obviously enjoyed it.'
36. Peter Eggeling interview.
37. Don Beard interview.
38. White sisters interview.
39. Bill, in Bill & Gill Pullen interview.
40. Ian Mullooly interview.
41. Ibid.
42. Alby Burney interview.
43. Christina, in Ben & Christina Morunga interview.
44. Ibid.
45. Arthur McRae interview.
46. Jack Curtis interview.
47. Edgar Jones, *Autobiography of an Early Settler in New Zealand*, Coulls Somerville Wilkie Ltd, Wellington, 1933, p. 47.
48. Samuel Butler, *A First Year in Canterbury Settlement*, edited by A. C. Brassington & P. B. Maling, Blackwood & Janet Paul, Hamilton, 1964, p. 52.
49. A. D. McIntosh (ed.), *Marlborough: A Provincial History*, Capper Press, Christchurch, 1977, pp. 102–103.
50. New Zealand Parliamentary Debate 24, 1877, p. 654, cited in Janet Holm, *Nothing but Grass and Wind: The Rutherfords of Canterbury*, Hazard Press, Christchurch, 1992, p. 17.
51. Margaret Wigley, *'Ready Money': The Life of William Robinson of Hill River, South Australia and Cheviot Hills, North Canterbury*, Canterbury University Press, Christchurch, 2006, p. 271.
52. A. Maud Moreland, *Through South Westland: A journey to the Haast and Mount Aspiring New Zealand*, Witherby & Co., London, 1911, pp. 77–78. This hut is now known as Blowfly Hut.
53. John Sullivan interview.
54. Laurie McVicar interview.
55. Carol MacKenzie interview.
56. Murray Dymock interview.
57. Ibid.
58. Bill & Gill Pullen interview.
59. Bill Traves interview.
60. Arthur McRae interview.
61. *Hawke's Bay Today*, Monday, 1 December 2008, p. 4.
62. Bill, in Bill & Gill Pullen interview.
63. Ibid.
64. Noel Martin interview.
65. Arthur McRae interview.
66. Bill, in Bill & Gill Pullen interview.
67. Arthur McRae interview.
68. Ray Stevens interview.
69. Ian Mullooly interview.
70. Arthur McRae interview.
71. Bill, in Bill & Gill Pullen interview.
72. Gill, in Bill & Gill Pullen interview.
73. Poem by an anonymous drover just before WWI, about the hospitality at Otairi Station, quoted in Philip Holden (ed.), *Station Country III: The Last Muster*, Hodder Moa Beckett, Auckland, 1997, p. 130.

9. The Twilight Years

1. Arthur McRae interview.
2. Stuart Barwood of Barwood Motors, Fairlie, telephone conversation, 21 March 2012.
3. By the early 1970s railways livestock traffic had all but disappeared. An attempt to revive the railing of livestock, in 1994, saw Tranz Rail experiment with carrying livestock in intermodal crates. However, this did not progress beyond the experimental stage as meat works had long stripped the railway sidings from their sites. Robin Bromby, *Rails That Built a Nation: An Encyclopedia of New Zealand Railways*, Grantham House, Wellington, 2003, p. 127.
4. Arthur McRae interview.
5. Guy Shanks interview by Helen Martin.
6. John Sullivan interview.
7. *New Zealand Farmer Stock and Station Journal*, July 1927, vol. XLVII, no. 7, p. 809.
8. Fred Cowin interview.
9. Bruce Ferguson interview.
10. John Sullivan interview.
11. Osborne, p. 27.
12. New Zealand Census of Population and Dwelling, Statistics New Zealand.
13. Statistics New Zealand website: www.stats.govt.nz/browse_for_stats/people_and_communities/geographic-areas/urban-rural-profile/historical-context.aspx
14. Owen Gibbens interview.
15. Arthur McRae interview.
16. David Stroud interview.
17. Barbara Kenton interview.
18. Sam Jefferis interview.
19. Arthur McRae interview.

20. Statistics New Zealand website: www.stats.govt.nz/browse_for_stats/people_and_communities/geographic-areas/urban-rural-profile/historical-context.aspx
21. Bill, in Bill & Gill Pullen interview.
22. Arthur McRae interview.
23. Ray Stevens in Sam Jefferis interview.
24. Guy Shanks interview by Helen Martin.
25. Arthur McRae interview.
26. Guy Shanks interview by Helen Martin.
27. *Waikato Times*, 4 Jan. 1975, p. 1.
28. Ibid., 23 Jan. 1975, p. 1.
29. Ibid., 7 Jan. 1975, p. 1.
30. Ibid., 19 Jan. 1975, p. 1.
31. Peter Cloake interview.
32. Ibid.
33. Ibid.

Epilogue
1. Captain Cook landed sheep in New Zealand in 1773; a ewe and a ram. They did not survive more than a few days.
2. Don Monk interview.
3. Murray Dymock interview.
4. Ian Mullooly interview.
5. Jack Curtis interview.
6. Laurie McVicar interview.
7. Jack Curtis interview.
8. Hugh Monahan interview.
9. Colin Bass interview.
10. Bill, in Bill & Gill Pullen interview.
11. Hugo Finn, 'The Song of the Drover', cited in A. E. Woodhouse (ed.) *New Zealand Farm and Station Verse 1850–1950*, Whitcombe & Tombs, Christchurch, 1950, p. 56.

GLOSSARY

boner stock Cattle that are past their breeding best and are taken to the works to become minced meat
box To mix mobs of stock
burn-off Any land thickly matted with matagouri and spaniard was set alight in order to clear land for ease of travel and for pastoralism; the new growth made good feed for stock
cast for age Breeding ewes that were no longer able to breed on hill country were bred on lower country for another few seasons
cross-breed Two breeds of animal bred to introduce a new, better breed more suited to the market
culled for age Similar to cast-for-age ewes; both cattle and sheep were farmed in this way
fat stock Stock fattened ready to be killed for consumption
hooler Someone who does not take care of stock on the road, rushing them rather than moving them quietly along – the cowboy element
knocking up Stock that are going lame
line Animals of uniform size, breed and formation were put into lines for sale
long acre The grassy roadside verge
matagouri Wild Irishman (*Discaria toumatou*) – a thorny, small-leaved bush or small tree found mostly in the South Island tussock country
new chum A name given to men who had just arrived in the country at the time of settlement: their dress and mannerisms gave away the fact they had not long been in the colony. Men were keen to rid themselves of such a title
pannikin A small metal pan or cup
punching Applying pressure to a mob to move them faster
quart pot Tin container for drinking, holding a quart of liquid
Rehab blocks Land developed for Returned Servicemen to settle on and farm
spaniard Speargrass (*Aciphylla*) – a sharp, spiny plant 1–3 metres tall, depending on the species
store stock Stock not yet ready to be killed for consumption – farmers with land unsuitable for fattening stock would sell their stores to farmers with flatter land where they would be fattened
terminal sire A meat ram or bull whose young are bred solely for the meat trade
tallow Rendered animal fat, used to make candles and soap
tonguing Animals that are overheated, distressed or exerting themselves are prone to tonguing: the tongue lolls out to the side
tutu or 'toot' *Coriaria arborea*, a native plant that is poisonous to stock
wagon box The body of a wagon
whip crackers The end of a whip that makes the cracking noise; often made from stripped flax tightly twisted

BIBLIOGRAPHY

Published

A History of Canterbury, vol. I, James Hight & C. R. Straubel (eds), Whitcombe & Tombs Ltd, Christchurch, 1957.

A History of Canterbury, vol. II, W. J. Gardner (ed.), Whitcombe & Tombs Ltd, Christchurch, 1971.

Alley, G. T. & D. O. W. Hall, *The Farmer in New Zealand*, Department of Internal Affairs, Wellington, 1941.

Amodeo, Colin, *The Mosquito Fleet of Canterbury: An Impression of the Years 1830–1870*, Caxton Press, Christchurch, 2005.

Bagnall, A. G., *Wairarapa: An Historical Excursion*, Masterton Trust Lands Trust, Masterton, 1976.

Bardsley, Dianne, *The Land Girls: In a Man's World, 1939–1946*, Otago University Press, Dunedin, 2000.

Bateman New Zealand Historical Atlas / Ko Papatuanuku e Takoto Nei, Department of Internal Affairs, Wellington, 1997.

Beattie, Herries, *Pioneer Recollections, Chiefly of the Mataura Valley*, Gore Publishing Co., Gore, 1911.

———, *Early Runholding in Otago*, Otago Daily Times and Witness Newspapers, Dunedin, 1947.

———, *The Southern Runs*, Gore District Historical Society, Gore, 1979.

Bidwill, William Edward & A. E. Woodhouse (eds), *Bidwill of Pihautea: The life of Charles Robert Bidwill: an account of the early days at Pihautea, New Zealand and a short history of the Bidwill family*, Coulls Somerville Wilkie Ltd, Christchurch, 1927.

Bradshaw, Julia, *The Far Downers: The People and History of Haast and Jackson Bay*, University of Otago Press, Dunedin, 2001.

———, *The Land of Doing Without: Davey Gunn of the Hollyford*, Canterbury University Press, Christchurch, 2007.

Bromby, Robin, *Rails That Built a Nation: An Encyclopedia of New Zealand Railways*, Grantham House, Wellington, 2003.

Brooking, Tom, 'Economic Transformation', in G. W. Rice (ed.), *The Oxford History of New Zealand*, 2nd edn, Oxford University Press, Auckland, 1992.

Buick, T. Lindsay, *Old Marlborough: Or, The Story of a Province*, 2nd edn, Capper Press, Christchurch, 1976.

Butler, Samuel, *A First Year in Canterbury Settlement*, edited by A. C. Brassington & P. B. Maling, Blackwood & Janet Paul, Hamilton, 1964.

Button, John, *Lyalldale: A Vision Realised*, Lyalldale Historical Group, Timaru, 2000.

Carmen, Eric, *The Broadwood Story 1888–1980*, 2nd edn, News Ltd, Kaikohe, 1983.

Carter, Bill & John MacGibbon, *Wool: A History of New Zealand's Wool Industry*, Ngaio Press, Wellington, 2003.

Chadwick, Tim, *Trucking Along: A Pictorial History of Trucks in New Zealand*, Grantham House, Wellington, 2001.

Churchman, Geoffrey B. & Tony Hurst, *The Railways of New Zealand: A Journey Through History*, Transpress, New Zealand, 2001.

———, *South Island Main Trunk*, IPL Books, Sydney, Wellington, 1992.

Clarke, Russell (ed.), *Kainui Story: A Profile of a Small Waikato Farming Community in the 1930s and '40s*, Hamilton, R. Clarke, c2006.

Collinson, Ossie, *Shouldering the Load: A Pictorial Tribute to Southland's Transport Pioneers*, Southland Road Transport Association, Invercargill, 1995.

Condliffe, J. B., *New Zealand in the Making: A Study of Economic and Social Development*, 2nd edn, George Allen & Unwin Ltd, London, 1959.

Corpe, Robyn, *Makino Memories*, Carrington Print, Feilding, 2001.

Crawford, Sheila S., *Sheep and Sheepmen of Canterbury 1850–1914*, Simpson & Williams, Christchurch, 1949.

Cross, Peter (ed.), *New Zealand Agriculture: A Story of the Past 150 Years*, NZ Rural Press Ltd, Auckland, 1990.

Dalley, Bronwyn & Gavin McLean (eds), *Frontier of Dreams: The Story of New Zealand*, Hodder Moa Beckett, Auckland, 2005.

Dalton, Clive, *Farm Dogs: Breeding, Training & Welfare*, NZ Rural Press Ltd, published in association with Dept of Agriculture, Waikato Polytechnic, Auckland, 1996.

Dawber, Carol & Cheryl Win, *North of Kahurangi, West of Golden Bay*, Picton, River Press, 2001.

———, *Ferntown to Farewell Spit*, Picton, River Press, 2003.

De Jardine, Margaret, *Little Ports of Taranaki: Being Awakino, Mokau, Tongaporutu, Urenui, Waitara, Opunake, Patea, together with some historical background to each*, M. de Jardine, New Plymouth, 1992.

Dew, Leslie, *The Tidal Travellers: The Small Ships of Canterbury,* A & M Publishers, Christchurch, 1991.

Duncan, Alfred H., *The Wakatipians; or Early Days in New Zealand,* Lakes District Centennial Museum Inc., Arrowtown, first published 1888, reprinted 1969.

Ffiske, Dave, *Lands & Survey, 1950 to 1965: Northland Scrub to Grass,* D. Ffiske, Queenstown, 2008.

Fraser, C. J., *New Zealand Trucks and Trucking,* Golden Press, Auckland, 1981.

Fulton, Myra, *Okuru: The Place of No Return: The Story about the History of Early Settlers of South Westland in New Zealand,* Myra Fulton, Takaka, 2004.

Gordon, Barrie, *Matawhero, Elsewhere and Other Things: A Livestock Auctioneer in New Zealand,* B. Gordon, Gisborne, 1990.

Graham, Jeanine, *Frederick Weld,* Auckland University Press/Oxford University Press, Auckland, 1983.

Gundry, Sheridan, *Making a Killing: A History of the Gisborne–East Coast Freezing Works Industry,* Tairawhiti Museum, Gisborne, 2004.

Hall, Sir John KCMG, 'Sheep-driving in the early days', *Canterbury Old and New 1850–1900: A Souvenir of the Jubilee,* Whitcombe & Tombs Ltd, Christchurch, c1900.

Harrison, Godfrey, *Borthwicks: A Century in the Meat Trade, 1863–1963,* Borthwicks, London, 1963.

Henderson, Jim, *Jim Henderson's Open Country,* Heinemann, Auckland, 1982.

Hill, R. D., 'Pastoralism in the Wairarapa, 1844–53', in R. F. Watters (ed.), *Land and Society in New Zealand: Essays in Historical Geography,* Reed, Wellington, 1965.

Holden, Philip (ed.), *Station Country III: The Last Muster,* Hodder Moa Beckett, Auckland, 1997.

Holm, Janet, *Nothing but Grass and Wind: The Rutherfords of Canterbury,* Hazard Press, Christchurch, 1992.

——, *Caught Mapping: The Life and Times of New Zealand's Early Surveyors,* Hazard Press, Christchurch, 2005.

Holmes, David, *My Seventy Years on the Chatham Islands,* Shoal Bay Press, Christchurch, 1993.

Hudson, Patrick, *Bridges of New Zealand,* IPL Books, Wellington, 1993.

Hunter, Brian & Rosalie Hunter, *A Collection of Stories Representing 125 Years of the Feilding Saleyards 1880–2005: The Smell of Success,* Feilding Information Centre, Feilding, 2005.

Hurst, Tony, *The Otago Central Railway: A Tribute,* Transpress, Wellington, 1990.

Hursthouse, Charles, *New Zealand the 'Britain of the South' with a chapter on the Native War, and our Future Native Policy,* Edward Stanford, London, 1861.

Johnson, David, *Auckland by the Sea: 100 Years of Work and Play,* David Bateman, Auckland, 1988.

Jones, Edgar, *Autobiography of an Early Settler in New Zealand,* Coulls Somerville Wilkie Ltd, Wellington, 1933.

Kennaway, Laurence J., *Crusts: A Settler's Fare Due South,* reprint, Capper Press, Christchurch, 1970.

Leary, Norman, *77 Years Among the Kowhais: Beyond Hunterville 1896–1973,* Service Printers Ltd, Wellington, 1977.

Leitch, David & Bob Stott, *New Zealand Railways: The First 125 Years,* Heinemann Reed, Auckland, 1988.

Leov, L. C. F., *As the Years Went By Between Greville and the Rai,* L. C. F. Leov, Blenheim, 1970.

Lovat, Lady Alice, *The Life of Sir Frederick Weld, G.C.M.G.: A Pioneer of Empire,* John Murray, London, 1914.

Lowe, David & Bryan Trim, *New Zealand's Cavalcade of Trucks. No.2. Kenworths and Macks,* Lodestar Press, Auckland, 1980.

Macgregor, Miriam, *Early Stations of Hawke's Bay,* A.H. & A.W. Reed, Wellington, 1970.

McCrystal, John, *100 Years of Motoring in New Zealand,* Hodder Moa Beckett, Auckland, 2003.

McIntosh, A. D. (ed.), *Marlborough: A Provincial History,* 2nd edn, Capper Press, Christchurch, 1977.

McIntyre, W. David (ed.), *The Journal of Henry Sewell 1853–7: February 1853–May 1854,* vol. 1, Whitcoulls, Christchurch, 1980.

——, *The Journal of Henry Sewell, 1853–7: May 1854–May 1857,* vol. II, Whitcoulls, Christchurch, 1980.

McLauchlan, Gordon, *The Farming of New Zealand: An Illustrated History of New Zealand Agriculture,* Australia & New Zealand Book Company, Auckland, 1981.

McLeod, George, *My Droving Days,* Ezyprint Solutions (self-published), Dunedin, 2011.

McNab, Robert (ed.), *Historical Records of New Zealand. Vol. 1,* Government Print, Wellington, 1908.

——, *Historical Records of New Zealand. Vol. 2,* Government Print, Wellington, 1914.

McQueen, Euan, *Rails in the Hinterland: New Zealand's Vanishing Railway Landscape,* Grantham House, Wellington, 2005.

Maling, Peter Bromley (ed.), *The Torlesse Papers: The Journals and Letters of Charles Obins Torlesse*

Concerning the Foundation of the Canterbury Settlement in New Zealand 1848–1851, 2nd edn, Caxton Press, Christchurch, 2003.

Martin, John E., *The Forgotten Worker: The Rural Wage Earner in Nineteenth-century New Zealand,* Allen & Unwin, Wellington, 1990.

Meyer, R. J., *All Aboard: The Ships and Trains That Served Lake Wakatipu,* New Zealand Railways & Locomotive Society, Wellington, 1980.

Millen, Julia, *Colonial Tears and Sweat: The Working Class in Nineteenth-century New Zealand,* Reed, Wellington, 1984.

Mincham, Carolyn, *The Horse in New Zealand: Attitude & Heart,* David Bateman, Auckland, 2011.

Moorhead, Murray, *Settler Tales of Old New Plymouth,* Zenith Publishing, New Plymouth, 2005.

Moreland, A. Maud, *Through South Westland. A Journey to the Haast and Mount Aspiring New Zealand,* Witherby & Co, London, 1911.

Nightingale, Tony, *White Collars and Gumboots: A History of the Ministry of Agriculture and Fisheries 1892–1992,* Dunmore Press, Palmerston North, 1992.

Nolan, W. D. (Bill), *The Droving Days: A History of Cattle Raising and Droving in South Westland,* W. D. Nolan, Christchurch, 1998.

Olssen, Erik & Maureen Hickey, *Class and Occupation: The New Zealand Reality,* Otago University Press, Dunedin, 2005.

Osborne, Sonny, *Droving Dogs and Sheep,* S. Osborne, Feilding, 1987.

Pickering, Mark, *The Southern Journey,* Mark Pickering, Christchurch, 1997.

Plummer, Peter, *From Drover to Director,* A. H. Plummer, Dannevirke, 1995.

Pullen, Gill, *On the Road Again: My Story of Life on the Road Droving with a Young Family,* Gill Pullen, Hunterville, 1996.

Rea, Jim, *Honey and Cheese,* J. Rea, Westland District, 1991.

Reeves, William Pember, *The Long White Cloud: Ao Tea Roa,* 3rd edn, Golden Press, Auckland, 1973.

Richards, Bill, *A Pioneer's Life,* Benton-Guy, Auckland, 1989.

Richards, E. C. (ed.), *Diary of E. R. Chudleigh 1862–1921, Chatham Islands,* Cadsonbury Publications, Christchurch, 2003.

Roxburgh, Irvine, *Wanaka Story: A History of the Wanaka, Hawea, Tarras, and Surrounding Districts,* Otago Centennial Historical Committee (Whitcombe & Tombs Ltd), Dunedin, 1957.

——, *Jacksons Bay: A Centennial History,* A.H. & A.W. Reed, Wellington, 1976.

Salisbury, Doug, *I Was So Lucky: Memories of 40 Years Working on Back Country Stations in New Zealand,* D. Salisbury, Taupo, 1994.

Simpson, Frank, *Chatham Exiles: Yesterday and To-day at the Chatham Islands,* Reed, Wellington, 1950.

Statistics of New Zealand for the Crown Colony. Period: 1840–1852. Dept of Economics, Auckland University College, Auckland, 1954.

Stephens, P. R., 'The Age of the Great Sheep Runs, 1856–80', in R. F. Watter (ed.), *Land and Society in New Zealand: Essays in Historical Geography,* Reed, Wellington, 1965.

Stevens, P. G., *Pyne, Gould, Guinness Ltd: The Jubilee History 1919–1969,* Pyne, Gould, Guinness Ltd, Christchurch, 1970.

Studholme., H., *The Addington Yards: A Century of Service to Farming,* Canterbury Saleyards Co. Ltd, Christchurch, 1975.

Sydney, Grahame, *Promised Land. From Dunedin to the Dunstan Goldfields,* Penguin Books, Auckland, 2009.

Tomlinson, J. E., *Remembered Trails,* J. E. Tomlinson, Nelson, 1968.

Tyrrell, A. R., *River Punts and Ferries of Southern New Zealand,* Otago Heritage Books, Dunedin North, 1996.

Vance, William, *High Endeavour: Story of the Mackenzie Country,* Reed, Wellington, 1980.

Watson, James, *LINKS: A History of Transport and New Zealand Society,* Ministry of Transport Te Manatu Waka, Wellington, 1996.

Wigley, Margaret, *'Ready Money': The Life of William Robinson of Hill River, South Australia and Cheviot Hills, North Canterbury,* Canterbury University Press, Christchurch, 2006.

Wolfe, Richard, *A Short History of Sheep in New Zealand,* Random House, Auckland, 2006.

Woodhouse, A. E., *George Rhodes of the Levels and His Brothers, Early Settlers of New Zealand: Particularly the Story of the Founding of the Levels, the First Sheep Station in South Canterbury,* Whitcombe & Tombs Ltd, Auckland, 1937.

Woodhouse, A. E. (ed.), *New Zealand Farm and Station Verse 1850–1950,* Whitcombe & Tombs, Christchurch, 1950.

Wright, Matthew, *Trucks Across New Zealand,* Whitcoulls, Auckland, 2006.

Wright, Shona, *Clifton: A Centennial History of Clifton County,* Clifton County Council, Waitara, 1989.

Wright, Vernon & Bruce Foster, *Stockman Country: A New Zealand Mustering Adventure,* Listener, Wellington, 1983.

Young, David, *Histories from the Whanganui River: Woven By Water*, Huia, Wellington, 1998.

Government publications

New Zealand Census of Population and Dwelling, Statistics New Zealand.

New Zealand Official Yearbooks (NZOYB), Statistics New Zealand.

Te Ara – Encyclopedia of New Zealand: teara.govt.nz

'Beef cattle and beef production', in A. H. McLintock (ed.) *An Encyclopaedia of New Zealand*, originally published in 1966, updated 22-Apr-09, www.TeAra.govt.nz/en/1966/beef-cattle-and-beef-production

'Stratification of the Industry', in A. H. McLintock (ed.), *An Encyclopaedia of New Zealand*, 1966, updated 23-Apr-09, www.TeAra.govt.nz/en/1966/lamb-and-mutton-production/3

'The Motor Age', in A. H. McLintock (ed.), *An Encyclopaedia of New Zealand*, 1966, updated 23-Apr-09, www.TeAra.govt.nz/en/1966/road-transport/2

Clark, Gary, Neville Grace & Ken Drew, 'Diseases of sheep, cattle and deer – Sheep diseases: worms, scab and anthrax', updated 1-Mar-09, www.TeAra.govt.nz/en/diseases-of-sheep-cattle-and-deer/2

Keane, Basil, 'Hoiho – horses and iwi: Horsemanship', updated 1-Mar-09, www.TeAra.govt.nz/en/hoiho-horses-and-iwi/3

Stringleman, Hugh & Robert Peden, 'Sheep farming', updated 11-Jul-13, www.TeAra.govt.nz/en/sheep-farming

Wassilieff, Maggy, 'Poisonous plants and fungi – Poisonous plants used for food', updated 1-Mar-09, www.TeAra.govt.nz/en/poisonous-plants-and-fungi/4

Journals and articles

Dalgety Digest, 'All's Fair in the Far North', 1980 Summer Edition, pp. 1–2.

Doubtless Bay Times, 'Early Beginnings: Dan Lewis', Kaye Dragicevich, 23 April 2008, p. 4.

——, 'Early Beginnings: Ken "Cassidy" Lewis', Kaye Dragicevich, 7 May 2008, p. 4.

New Zealand Gazette and Wellington Spectator, vol. IV, issue 282, 20 September 1843, p. 2.

NewsPlus, 'Men of the Road', Murray Patterson, 23 May 1990, pp. 8–9.

——, 'Out in the gig with Len Gilbert', Murray Patterson, 13 June 1990, pp. 10–11.

New Zealand Herald, 'Last Far North droving trek seems just around the bend', Ross Barrett, 22 September 1979, p. 12.

New Zealand Woman's Weekly, 'She runs her own farm and lonely airport', Garth Gilmore, 17 June 1963.

——, 'Hard life but he loves it', 2 May 1978.

Northern Advocate, 'A lifetime droving in North', Susan Botting, 25 October 1984, pp. 4, 7.

Rodney's Rural Lifestyle, Martin, Helen, 'Guy Shanks Drover', November 2010, issue 38, pp. 9, 10, 19.

The Dominion, 'Drover's life only for the patient', 5 June 1981, p. 7.

Waikato Times, '4000 sheep begin "trek" to Horotiu', 3 January 1975, p. 1.

——, 'Traffic men may halt big sheep drive', 4 January 1975, p. 1.

——, 'Sheep will be diverted from main highway', 7 January 1975, p. 1.

——, 'Legal steps to control drives to be taken', 10 January 1975, p. 1.

——, 'Drover quits sheep trek as losses mount', 15 January 1975, p. 1.

——, 'Mob to be checked to see farmers keep assurances', 17 January 1975, p. 1.

——, 'End of the road for 3400 ewes', 23 January 1975, p.1.

——, 'Caution urged in moving cattle herds on roads', 31 May 1989, p. 5.

Hawkes Bay Today, 'Remember our drovers killed in war', Lawrence Gullery, 1 December 2008, p.4.

Unpublished

Beard, Trevor E. W., 'Memoirs 1907–2000', Feilding.

Bredin, Lial, 'Droving on the Coast', New Plymouth.

——, 'Schoolboy Drovers', New Plymouth.

Davies Family, 'Some Recollections of Our Father: A Drover and Droving in Its Heyday', Feilding, 2002.

MacKenzie, Carol, 'Memories of Droving Days', for the Tainui Historical Society Museum, Mokau, Taranaki.

Morgans, Ivor, 'Droving in Southern Hawkes Bay', Dannevirke.

Richards, Harry, 'Richards Family History', 2013.

Stephenson, Robert J., 'Reminiscences, West Coast, 1930', TSO Historic Resources, West Coast Conservancy, Department of Conservation.

Swan, Lew, 'Droving Day: A Recollection of Stock Droving in Southern Hawkes Bay 1941–1952', Dannevirke.

Alexander Turnbull Library (ATL)

Chapman, Alfred, Journal from Hawke's Bay to Wellington, 4 Jan–18 Feb 1856, qMS-0415.

Clayton Station Diary, 1895, qMS-0472.

Evans, Roberts Joseph, Farm Diary, 1939, qMS 0689; 1944, qMS-0693; 1945, qMS-0694; 1947, qMS-0696; 1948, qMS-0697; 1950, qMS-0699; 1952, qMS-0701; 1954, qMS-0704.

Hamilton, Francis William, 1840–1901, Outward Letters 1861–1862, MS-Papers-4328.

Heays, W. H., Diary, transcribed, ca1923–1924, MS-Papers-9009-19.

Jollie, Edward, '1825–1894 Reminiscences', qms-1072.

Te Awaiti diaries, transcribed, 1863–1890, MS_Papers-6567-1.

Tiffen, F. J., Journal. A record of significant events, 1845–1911, MS-Papers-1348-01.

Canterbury Museum

Hewson, A., 'Reminiscences: Early days in the Ashburton County', ca1850s-1890s. ARC 1900.257.

McGregor, John, 'Reminiscences: A short sketch of my life', 1863, ARC 1900.337.

Hocken Collections, Uare Taoka o Hakena, University of Otago

Notes on Otago Central in 1857 and 1858, MS-582/D/5/c.

Macmillan Brown Library, Te Puna Rakahau o Macmillan Brown

Acland Family papers, MB44, Box 52, 56, 59, 84; MB44B2/41, Box 7.

Canterbury Frozen Meat Company records, MB1660, Box 117, 127, 128, 161.

Pyne Guinness Corporation records, MB1990, Box 61.

Marlborough Museum

Hill, Martin, Map of early stock routes in Marlborough.

Murchison Museum

Peacock, Alma, 'To Market', Journal 8.

South Canterbury Museum

Annett, William, Farm Diary Transcripts 1890-1958 Genealogical notes (8pp), part of 2003/020.1.

Denne, Mark, *South Canterbury Railways: Index to Timaru Herald references.*

Teschemaker, W. H., Diary Kauro Hill Station Dec. 1855–Feb. 1864, 27/2.

Theses

Crawford, Sheila S., 'Sheep and their place in the development of Canterbury', MA thesis, University of New Zealand, 1944.

Hargreaves, R. P., 'Speed the plough: An historical geography of New Zealand farming before the introduction of refrigeration', PhD thesis, University of Otago, 1966.

Kelly, Clare M., 'One day to the rainbow: The accommodation houses on the Nelson to Canterbury inland stock route, 1855–1900', MA thesis, University of Auckland, 2007.

Moore, Rae L., 'The history of Cheviot Hills 1846–1893', MA(Hons) thesis, University of New Zealand, 1937.

Peden, Robert, 'Sheep farming practice in colonial Canterbury 1843 to 1882: The origin and diffusion of ideas, skills, techniques and technology in the creation of the pastoral system', MA thesis, University of Canterbury, 2002.

Ruth Low Personal Correspondence

Aitken, Laurel, material on uncle, 2007.
Allan, Aveley, material on grandfather, 2007.
Barnes, Glad, email, 2006.
Bell, Peter, email, 2006.
Butland, Joan, letter, 2007.
Douglas Smales, Veller, email, 2006.
Esler, Sam & Dot, email, 2006.
Gilbert, Roger, letter and notes, 2009.
Grieve, Angela, email, 2006.
Hill, Billy, notes.
Johnson's Barge Service Ltd, email, 2013.
Leitch, Selwyn, email, 2006.
Leov, Faye, email, 2008.
Lusty, Margaret, letter and photo, 2007.
McLean, Doris, email, 2006.
Mabey, Helen, letter and photos, 2008.
Main, William, letters, 2007.
Maisey, Eva, letter, 2007.
Mathias, Barbara, letter, 2007.
Munro, Gavin, email, 2006.
Nelley, Iris, letter and notes.
Patterson, Geoff, email, 2006.
Percy, Max, email, 2006.
Quirk, Jocelyn, letter and account of father's droving.
Schmidt, Linda (née Osborne), reminiscences.
Simpson, June, letter, 2003.
Sowerby, Julia, emails and photos, 2004.
Stimpson, Lennie, letter, 2007.
Weenink, Julia, letter, 2006.
Williams, Barbara, email, 2007.
Win, Cheryl, letter, 2008.
Win, Rodger, email, 2007.
Wright, Shona, email, 2007.

ACKNOWLEDGEMENTS

A serendipitous meeting drew me into the world of droving. I was a town dweller with no farming background; I had never given a thought to the role of the drover in our country's history. Armed with minimal knowledge but a great deal of curiosity and enthusiasm, I started this oral history project by interviewing 10 individuals who were part of the droving story in and around the pastoral hub of Feilding. My curiosity was piqued, I could not stop at just 10 interviews – and so my immersion in New Zealand's droving history really began. As is apt to happen when one places no parameters on a project, it grew arms and legs and ran away with me. After some 60 interviews it was well and truly time to rein in the beast and begin to meld the drovers' words into a book. I hope as you turn these pages their voices resonate. To those men and women who willingly opened their homes and their lives to me, I say a profound thank you. Without you there would be no book.

Various newspapers obligingly put out SOSs for information and names; Jo Keppel of the *West Coast Messenger*, in particular, was very supportive. As a result many people emailed, sent letters and photos. Not all of the stories could be included, nor everyone interviewed, but to all of you who took the time to make contact, please know that your stories added to my knowledge and understanding of this subject and helped create the backdrop to my writing.

Along with the men and women interviewed there are many individuals and organisations to acknowledge and thank. Impetus for the oral history research came from funding through two grants from the Ministry for Culture and Heritage's Australian Sesquicentennial Gift Trust Oral History Award. I am most grateful to the Ministry for this funding and also for the encouragement, advice and support of Megan Hutching, the Senior Oral Historian at the time. Further research was also funded by the Ministry's New Zealand History Research Trust – thank you for believing that the work I had begun was worth continuing. Thank you also to all those at the Oral History Centre at the Alexander Turnbull Library, in particular Linda Evans and Gillian Headifen, for your assistance and advice.

New Zealand is well endowed with wonderful museums and libraries, many of which I have used throughout my research. I would like to acknowledge the assistance of the Alexander Turnbull Library, Hocken Library, Archives New Zealand, New Plymouth's Puke Ariki, Canterbury Museum, Nelson Provincial Museum, Marlborough Museum, Feilding Library, Palmerston North Library, Opotiki Museum, Murchison Museum, Tairawhiti Museum in Gisborne, Hokitika Museum, and the Museum of Transport & Technology. To all the wonderful librarians and archivists I have hounded, a huge thank you. Special mention must be made of the help given by Erin Kimber of the Macmillan Brown Library and Tony Rippin at the South Canterbury Museum. To Lynley Simmons at Timaru Public Library, thank you for readily sourcing a stream of books for me from around the country. Thank you also to Lynda Hammond of the Far North Museum for offering assistance and guidance when I was researching in the region. I would also like to thank Jackie Breen from the Department of Conservation for her assistance.

My heartfelt thanks to all those who assisted with maps and imagery, including the kind souls who helped with scanning and emailing towards the end of the book's production.

A generous grant by the Canterbury History Foundation and mentoring by Foundation members Dr John Cookson and Graeme Dunstall was the stimulus I needed to write. I am very grateful to both for providing invaluable expertise, gentle guidance, wisdom and encouragement. The trust and support from the Foundation, even when deadlines lapsed, has been greatly appreciated and I am in their debt.

As a first-time author the trepidations of being published have been greatly eased by the expertise and professionalism of the Penguin team. A huge thank you goes to commissioning editor Jeremy Sherlock, who from the start has shared my enthusiasm for this story and encouraged and cajoled and now seen the project through to its completion. Thanks must also go to Leanne McGregor for her incredible eye for detail and for ensuring each step of the production process went smoothly; and to Jenny Haslimeier for her wonderful design skills.

The refining process is never without pain so sincere thanks to my editor Gillian Tewsley for making this process as pain-free as possible. Thanks also to Jan Kelly for patiently untangling a maze of information in order to create the maps.

I would also like to acknowledge Don Beard and the late Audrey Severinsen of Feilding – the two people who started me on this journey. I am so glad we met and you encouraged me along this path.

The support of friends and family throughout the years as I interviewed, collected, collated and wrote have sustained and encouraged me. The book is *finally* finished, and you have each played a part in ensuring its completion. To my parents, Patricia and Stewart Entwistle, thank you for valuing education and encouraging an enquiring mind. To my two girls, Charlotte and Kate – you have been adventurers alongside me all the way, two gorgeous cheerleaders on the sidelines, and I love you for it. I am not sure how I begin to thank my husband without making him sound like a saint. Not wanting to give the wrong impression, I will not list all his virtues here. Suffice to say, Mark, you are my rock!

Ruth Entwistle Low
February 2014

INDEX

Bold entries refer to images.

accommodation 76, 125, 143, 148–49, 205, 211–19
Acheron River 24, 25
Ada River 185
Addington saleyards 53–54, 165, 169
Ahaura River 102
Ahuriri River 111
Aiken, Charlie 195
Allen, Bill 141
Amuri 128
Anderson, Peter 134
animal husbandry 131–35, 172–75, 194–97
Antrim (paddle steamer) 66–67
Aorere Bridge 155
Appleby 72, 88, 89
Arawhata River 72, 73–74, 75
Arrow River 30, 31
Arrowtown 138
Arthur's Pass 56, 171
Ashhurst 112, 145, 147, 231
Atiamuri **126**
Atwell, Jimmy (Jimmy Eh!) 124
Awatere River 218–19

Bagnall, A. G. 17
Ball, Ray 55
Banks Peninsula 215
Bannockburn 28
Barber, Allan 50, 68, 116, 140–41; biography 244
Bardsley, Dianne 106; *The Land Girls* 106–107
Barefell Pass 23, 24
barging stock **insert 2 p. 2**
Barnett, Mark 103, 241
Barwood Motors **226**, **insert 4 p.4**
Bass, Colin 236; biography 244
Bateman, Alec **226**
beaches, as stock routes 19, 20, 41, 67, 68, 72, 76, **82**, 85, 89, **124**, **136**, **137**, **160**, 181–82, **insert 2 p.1**
Beard, Don 182, 210; biography 239
Beattie, Herries 104; *Early Runholding in Otago* 153
Beazley, Dave 196
Beddingfield, Townie 146
Belfast freezing works 45, 111

Belgrave, Bob 244
Bell, Len 95, 124
Ben Lomond (twin screw steamship) 66
Bergman, Sarah 214
Berry, John 23, 24–26
bicycle, droving on 186
Bidwill, Charles 17, 21, 202–203
billy, boiling **97**, 109, 123, 127, 136, 206–**207**, 208, 209, 210, 218
Birchwood Station 111
Blue (Maori) River 158
Blue (Moeraki) River 213
boredom 105, 125, 235; *see also* monotony
Borrell, Les **226**
Borthwicks freezing works 46, 111, 145
Borwick, Bert **54**
Botherway, Percy 217
Bradley, Jessie 210, 248
Branch River 153–54
Bredin, Lial 129; biography 239
bridges 63, 146, 154–**59**, **insert 3 p. 2**; *see also* rivers
Broadwood 82, 105, 114; gathering in 2008 244; sale 82
Brown, Barney **54**
Buchanan family 76
bullocks/bulls, droving 43, 44, 51–53, 74, 75, 130, 168–69
Burney, Alby 6, 95; biography 239
Burnside freezing works 45, **46**
Burnside saleyards 51–53
Burr, Harry 28, 29
Burton, Kath 47, 110, 119, 196; biography 243
Burwood Station 54
Butler, Samuel 173, 211–12

Cairnmuir Range 29
Cameron, Archie 28
Cameron family 111
Campbell, Linc 95, 171
Campbell Island 68
camping 148–49
Canterbury: 211–12, 222; inland route to 22–27; map 18; Plains 22–27, 32, 107, 187; winds 32, 216

Canterbury Provincial Council 22
Cantwell, Dan 6
Cape Campbell 23
caravans 218–19, **insert 4 p. 3**
Carmen, Eric 114
Caroline Bay 111
Carterton 22
Cascade River 72, 73–74
Cassie, Jack **226**
cattle 40, 44–45, 73–74, 127, 129–30, 144–45, 226–27, 233; crossing rivers 144–45, **154**, **158**; numbers 40, 44, 45, 82
Central Plateau 58–59, 232; Maori drovers on 104
Chaney, Les **54**, 144
Chapman, Alfred **26**
Chasm Creek 78–79
Chatham Islands 68–**70**
Cheviot Hills estate 99
child drovers 96, 100–14, **101**, 179, 187
Christchurch 165, 168
Clarence River 24, 25, 27
Clark, Mickey 78, **202**; biography 244
Clark, Tom 113
Clements, Eric 118
Clements, Phyllis 83, 118–19, 162; biography 239
Cloake, Peter 48–49, 57, 229, 232–33; biography 240
clothing 9, 201–205
Clotworthy, Neville 85–86
Clyde 28
Coal Creek 28
Coal Creek Station, drove from 28
coat, oilskin **202**, 203, **insert 3 p. 4**
Cocks, Alex 95, 124, 161, 175; biography 244
Collingwood 88, 89, 155, 157
Colyton 112
Condon, Billy 172
Condon, Jack 142
Condon, Lewis **202**
Condon, Tom 80–81, **202**
Condon, Tony 73–74, 76, 117, 203; biography 245
Coogan, Pat insert 4 p. 3
cookbook 208–209
cooking 205–11

Coomerang (paddle steamer) 67–68
Cooper's dip 206, 211, **insert 3 p. 4**
Copper Creek 76
Corfield, Jim **54**
Coulter, Kerry 84
council bylaws 225, 231–32
country stores 206
Cowan, Bernie 78
Cowan family 76
Cowin, Fred 155, 188, 229; biography 245
Crawford, Allan 84, 169, **insert 3 p. 2, insert 4 p. 1**; biography 240
Crawford, Elisabeth 240
Cron, Adam 116
Cron, Allan 74–75, 76, 79, 116–17; biography 245
Cross Creek Station **43**
Crown Range 30
Curtis, Jack 'Boy' 47–48, 95, 109, **115**–16, 119, 127, 129, 132, 141, 156, 180–81, 188, 189, 194–95, 196, 197, 206, 211, 216; biography 240
Curtis, Jack Sr 115

daily progress 139–40; *see also* distances; duration
dairying industry 44; and droving 44, 136; sharemilkers' herds 44, 136
Dalgety's 48, 193, 229, 232
Davies, George 95
Deans, William 99
De-Arth, Jimmy 209
demise of droving 9, 222–34
Dillon, Spencer 127, 129, 194; biography 240
diseases, sheep 34
distances, of droves 17, 24, 31, 81, 104, 131
dogs, drover's 9, 11, 47, 71, 76, 116–17, 118, 143, 161, 179, 187–97; border collies 190–91; breeds 190–94; caring for 194–97; cattle dogs 190; food (tucker) 47, 86, 87, 123, 135, 145–47, 185, 195, 210; handy dogs 190–94; huntaways 191, 193; leading dogs 145, 164, 190, 192–94; training 187–90; wild 34

Donaldson, David 140–41
Dowell, Max 158–59, 209; biography 245
drafting 47–48
drinking 143, 144, 148, 212
drovers/droving 92–119; accommodation 211–19; animal husbandry 131–35, 172–75; avoiding trouble 125–26, 140–42, 161–62; backgrounds 95–98; boredom 105, 125, 235; camping 148–49; children 100–14; clothing 201–205; cooking 205–11; craft of 120–49; daily progress 139–40; demise of 9, 222–34; drinking 144, 148; economic role 36–59; food 10, 205–11; gentleman 98; learning 128–29; local knowledge 136–39; Maori 104, 113–14, 130, 141, 142; monotony 15–16, 79, 101, 102, 125, 234; night vigils 32; numbers 108–109, 114, 229; professionals 108–109; requirements 96, 98–101; routines 123–26, 139–40, 143; rural workers 98–100; self-perception 11; skills 122–43, 149, stock preferences 127, 129–30; stock training 127–30; storytelling 11, 80, 197, 214; trials of 31–33; twilight years 223–34; women 10, 96, 104–108; working hours 87
Drovers Memorial Huts **217**
droves, early ('golden fleece' age) 12–35, 38; danger to sheep 15, 33–35; Nelson to Canterbury 16, 22–27; Rees run, Lake Wakatipu 28–31, **insert 1 p.1**; trials of droving 31–33; Wairarapa, first drove into 16, 17–22
droves, iconic 71–91: Ken Lewis Far North drove 81–87; Ned and Jack Richards's drove 87–91; South Westland drove 72–81
droving cf. mustering 9
droving routes 15, 60–91; *see also* stock routes
drowning 153–59
Duncan, Alfred 16, 28, 29, 30, 31, 98–99; *The Wakatipians*

or Early Days in New Zealand 28, 98–99
Duncan, Ken 145, 147
Duncan family (Otairi Station) 55, 57, 112
Dunedin (ship) 40
Dunstan Range 30
duration, of droves 71, 81, 104, 114, 139–40, 149; *see also* distances
D'Urville Island 68, 88
dust 132
Dymock, Murray 169, 181, 192–93, 208, 209; biography 245

Earnslaw (twin screw steamship) 66, 68
East Coast drovers/droving 64–65, 104, 112–13, **124**, 148, 227, 234; Maori 104, 206
East Takaka 89
economy, drover's contribution to 36–59, 234
Eggeling, Betty 74, 75, 76, 79–80, 107, 135, 203, 214; biography 245
Eggeling, Charlie 74, 75, 76, 80, **insert 1 p. 4**; biography 245
Eggeling, Dick 159
Eggeling, Peter 209
Eggeling family 36
Elder brothers 128
environment/terrain 20, 22, 25, 26–27, 28, 72, 85, 96, 98; *see also* hardships; vegetation; weather
Esler, Jack 95, 128–29, 138–39
Esler, Sam 95

Fairview Station 194
families, droving 76, 108
Far North 41, 113–14, 118–19; Ken Lewis drove 60, 81–87; Maori drovers 104, 113–14; map 73
farming practices 40–45; intensification 40, 44–45, 225–26
fat stock buyer 45–47
Feilding 6, 46, 48, 49–50, 92, 96, 98, 103, 107, 109, 110, 112, 229, 230, 233; sale day 234; saleyards 48–51, **49, 50**, 57, 64, 110; statue of drover **92**

Ferguson, Bruce 141, **160**, 193–94, 195, 229; biography 245
Ferguson, Jimmy 138
Ferntown 88, 89
ferries 21, 65, 69, 212
Finch, Neil 244
Finn, Hugo 237
Fisherman's Paddock **insert 3 p. 1**
Fitzgerald, Tom 48
Fitzroy 102–103
Flaxbourne Station 23
Flower, Jack 62, 141, 195
food, drovers' 10, 205–11
Ford, Henry 16
Forrest Ford 54
Fox Glacier 76
Fox Glacier Hotel **214**
Fox River 76
Frank, Bev 188
Frank, Harry 144, **183**, 184, 186, 188; biography 240; his dog team **189**
Frasertown Bridge 155
freezing works 44, 45–47, **45**, **46**; and droving 46
French, Bob **54**
frozen meat trade 39–47
Fulton, Myra 78, 117–18; biography 246
Furniss, Bim 97, 156; biography 240
Fyers, Buckley 241
Fyers, Lee 194, 203; biography 241

Gael (motor vessel) **74**
gentleman drovers 98
Gibb, Helen 213
Gibbens, Dick 34, 170, **insert 3 p. 3**
Gibbens, Owen 71, 170, 196, 229; biography 246
gigs **176**, **186**
Gilbert, John 28
Gillies, Mr 86–87
Gisborne 112–13; drovers 64–65, **108**; sale 115
Glenhope 186
glossary 258
'golden fleece' 38, 12–35
goldrush, West Coast 56–57; and droving 56–57, 72
Gordon, Barrie 58–59, 64
Grace, Chris 113
Grace, Ike 142

Graham & Gebbie **198**
Grampian Hills Station 16
Grant, Colonel 29
Grant family 111
Grayling, Malcolm 102–103, 135; biography 241
Grayling, Phillip 102–103, 241
Great Barrier Island 68–69, **insert 2 p. 2**
Grey, Arnold 183
Grey, George 22
Grey River 185
Greymouth 71, 170
Grieve, Arthur 95
Grieve, Elliott 95
Grieve, Minnie 215–16
Grooby, Barry 89, 179; biography 246
Grooby, Lance 89, 246
grooming, personal 203–205
Gunn, Davey 183, 194
Gypsy Day 44

Haast River 72, 76, 116
hacks 180–83
Hall, John 100
Hamilton, Francis 172
Hanmer **126**
Hanmer Plain 24, 26–27
Harding, Ken 95, 206
hardships 10, 17–18, 30, 31–33, 76
Harihari 172
Harpers Pass 56–57
Harris family 36, 76, 79
Harrison, Keith 53–54
Harvey, Simon 28
Havelock 68
Havelock North 217
Hawai Beach **insert 2 p. 3**
Hawke's Bay 106, 112, 217; droving in 106, 231; pastoralism in 22
Hawkins, Kathleen 189
Hay, John 104
Heays, W. H. 67–68
Herekino Gorge 82, 162
heroic phase of droving 12–35
Heveldt, Johnny 203
hill country farms 40
Hobson, Albury 86–87
Hokitika 72
holding paddocks 125, 136, 140, 142, 143, 146–47, 192, 215–16, 229

Holland, John 86–87, 179, 182
Holland, Mike 174
Holland, Sidney 157
Holley, Billy 138
Hook, Tommy 203
hooler 94, 140, 149, 258
Horotiu freezing works 232
horses, drovers' 17, 23, 28, 73–74, 76, 104, 143, 176–87; drovers' preferences 180–81; feeding 184–85; and gigs **176**, **186**; hacks 180–83; packhorses 180, 183–85, 206
Houhora 82
hours, working 87
Hukatere **84**, 85
Hunter, Isabel 247
Hunterville 112
Huntly 156, 162; road-and-rail bridge 156
Hurley, Andy and Mike, biography 241
Hurley family (Papanui) 55, 57, 184
Hurunui River, drove from 25
Husband, Bill 95
Husband, Max 95

Inglewood **insert 1 p. 2**
interviews 239–48
island farms 68–71

Jackson, Jack 'Shotgun' 124
Jacksons Bay Special Settlement 72, 116
Jacobs River 76
Jefferis, Jack 55, 241, **insert 2 p. 3**
Jefferis, Sam 55, 126, 148; biography 241
Jenkins, Ken 247
Jollie, Edward 16, drove to Canterbury 22–27; his horse 23
Jollies Pass 24, **insert 1 p. 1**
Jones, Edgar 56–57, 128, 211
Jones, Karl 64, 134–35; biography 246

Kaihoka 229
Kaihu district **63**
Kaikohe 83
Kaikoura 23, 176
Kaimai Range 207
Kaitaia 82

Kakapotahi 214
Kane, Jack 95
Karamea 64, 134
Karangarua 76
Kauri saleyards 81, 83
Kawarau River 28, 29
Ken Lewis Far North drove 60, 81–87, 97, 162; map 83
Kennaway, Laurence 25, 33; *Crusts* **25**, 33, **134**
Kennedy, Eric 80; biography 246
Kennedy, Mary 80; biography 246
Kennedy, Pat 68, 155; biography 246
Kenton, Barbara 103; biography 246
King, Ralph 248
King Country 58–59, 129, 233
Kirkley, Wilfred 217
Kopungarara 17, 21

Lake Mapourika 76
Lake Onoke 20–21
Lake Paringa 76
Lake Station 153
Lake Wakatipu 16, 28, 31, **66**, **67**, 68, 138, 244
Lake Wanaka **36**
lamb trade 40
Lambert, Frank 123, 171, 208
land girls 10, 105–**106**
Lands and Survey blocks 58–59, 194–95; and droving 58–59, 104, 114
Landsborough run 116
laundry 204–205
leader of stock 140–41, 162
learning droving skills 128–29, 165
Lee, Edward 16, drove to Canterbury 22–27
Leslie, June 107; biography 247
Levels Station 27, 33
Levin saleyards **238**
Lewis Pass 183
Lewis, Ken, 118–19, 174, 181; drove 60, 81–87, 118–19, 141–42, 181, **196**, 231
Liberal Government 40
lifestyle blocks 230
Lindis Pass 222, **insert 4 pp. 3, 4**
'local' drovers 64–65
local knowledge 136–39
loneliness 101, 102, 196
long acre (road verges) 55, 57, 81

Lorneville saleyards 183
Low, Andrew 28, 30
Lower Moutere 91, 168

MacDonald, Captain 55, 95
MacKenzie, Carol **101**, 136, 215; biography 241
Mackenzie, James 27
Mackenzie Country 16, 111, 205
McClunnen, Joe 236
McCready, Bill 87
McCready, Selwyn 87
McCreath, Barbara 104
McGregor, John 16, 205
McKay, Dick 134
McLaren, Joe 95, 183
McLennan, John **215**
McLeod, George 51–53; *My Droving Days* 51–53
McNab, Alexander 153
McPherson family 76, 79
McRae, Arthur 57, 65, 95, 109, 112–13, 125–26, 131–32, 135, 140, 149, 155, 184–85, 188, 206, 216, 218, 230, 231, **insert 2 p. 4**; biography 241
McRoberts, Jock 55
McVern, Eddy 55
McVicar, Laurie 50–51, 102–103, 128, 129–30, 134, 138–39, 157, 171–72, 179–80, 181, 185–87, 203, 214–15, 235; biography 247
Mabey family insert 2 p. 2
Macky, Ned 141, 156
Manawatu district 64, 107, 184, 207, 229
Mangakahia River 156
Mangamuka Gorge 82, 83, 85
Mangaorapa 106
Maniototo Plains 28
Manuherikia 30
Maori: drovers/droving 104, 113–14, **130**, 141, 142; land purchased from 40; leases from 17; Maori Affairs land 114; negotiations with 21; providing food 206
Maori Saddle 41, 76, 78–79, 203
maps: Far North drove (Ken Lewis) 83; historic droves (Wairarapa, Wairau, Otago) 18; South Westland drove (Far Downers) 73; Tasman drove (Ned and Jack Richards) 88

Maraekakaho 217, **insert 3 p. 2**
Marlborough Sounds **69**
Marsden 235
Martin, John E. 99; *The Forgotten Worker* 99
Martin, Noel 164, 171, 191; biography 241
Maruia Hotel **215**
matagouri 16, 24, 258
Matahina 135, 138, 148, 162–63
Matamau 146
Matangi **57**
Mataura River 153
Matawhero Ford **108**
Matawhero saleyards 64
Matthews, Bronwyn 103–104
Matthews, June **106**
Matthews, Neil 103–104, 144; biography 242
May, Colin 144
merinos 17, 34, 39
Mitimiti 105
Moeraki River 76
Moerewa freezing works 81, 114, 196
Mokau River **154**
Molyneux (Clutha) River 28, 29, 30
Monahan, Hugh 180, 197, 208, 236; biography 247
Monk, Don 43, 53–**54**, 131, 144, 157, 174, 178, 181, 193, 235; biography 247
monotony 15–16, 79, 101, 102, 125, 234; *see also* boredom
Moreland, Maud 213
Morgan, Kath 89–91; biography 247
Morley's Pass (Thomson's Gorge) 30
Morrison, William 99
Morunga, Ben 105, 109, 113–14, 211; biography 242
Morunga, Christina 105, 211; biography 242
Morunga, Tom (Tom Clark) 113
Moses, Mo 85, 142
motorists *see* traffic
Motu Gorge 206
Motunau 213
Mouka Mouka (Mukamuka) Rocks 19–20
Mountaineer (paddle steamer) 66
Moyes, Scottie 95
Mt Peel Station 99–100

Muir, Jack 244
Mukamuka Rocks 19–20
Mullooly, Ian 59, 95, 124–25, 127, 187, 209; biography 242
Munro family 222, insert 4 pp. 3, 4
Murray, Dollar 97, 162
Murray, Rob 97, 181
mustering 9, 73, 85; cf. droving 9

Nalder, Doris 248
necessities of droving 17, 198–221
Ned and Jack Richards's drove 87–91; map 88
Nelson 22; droving from 16, 22–27
Nelson region 64, 88–89, 159
Nevis Creek 28
New Plymouth 103, 165
New Zealand Company 17
New Zealand Refrigerating Company 45
Newton, Thomas 153
Ngata, Apirana 104
Ngunguru 118
night vigils 32
Ninety Mile Beach **82**, 85, 119, **196**
Nolan, Bill 73–74, 76, 79; poem 77
Nolan, Des 73–74, 76, 78; *The Far Downers* 78
Nolan, Eddie **202**
Nolan, Kevin **202**
Nolan, Paddy 79
Nolan, Steve **202**
Nolans' runs 73–74
Norsewood 146
Northland 58–59, 158, 211
numbers of drovers 108–109, 114, 229

Oamaru **166**
O'Brien, Charlie 186
Okaihau 82
Okuru 72, 76, 117
Oliver, James 217
Opotiki 115
Orari Station 201
Osborne, Sonny 95, 96, 229; *Droving Dogs and Sheep* 229
Otago 28–31; drove map 18; runs 30
Otairi Station 55, 57, 112, 113, 135, 220–21
Otamarakau 132
Otangaroa 82

Otira Gorge **60**, 171
'overseas' drovers 64–65
Oxford 132
Oxnam, Gertrude 248

packhorses 180, 183–85, 206
Pakawau 89
Pakowhai **12**
pannikins 17, 30, 212, 258
Papanui Station 55
Paparoa (Northland) 84
Parakao 83
Parapara River 157–58
Pareora freezing works **45**
Paringa 72
Parkinson, W. L. (Bill/'Parkie') 53–55, 144
pastoral leases 17, 40
pastoral regions 16, 17
pastoralism and the economy 36–59, 225
Patons Rock Beach **insert 2 p. 1**
Paturau 41, 87, 89, 229
Paua 81, 82; sale 82
Percival, Augustus 23, 24–27
Peria, sale 82
Petre, Henry 17, 19
Philps, Brian 102, 145–47; biography 242
Pigeon Bay 104
pigs, droving **130**
Pipiwai–Mangakahia route 83
plants *see* vegetation
Pleasant Point sales 111
poems: 'The Drover' 8; 'The Drover's Dog' 189; 'A Recognition' 220–21; 'The Song of the Drover' 237
Pourere 22
Povey, Billy 167
Pratt, Myrtle 116–18
problems: encountered 10, 15, 17, 20, 21, 22, 31–33, 150–75; river crossings 15, 23, 24, 28–29, 30, 31, 32–33, **65**, 74–75, 76, 128, **150**, 153–59, **154**, **158**; in towns 165–72; traffic 160–65, **163**, **167**; tutu poisoning 172–75; *see also* avoiding trouble; hardships
professional drovers 108–109
provisions 10, 19, 23, 205–11; *see also* cooking; food
Pukehamoamoa 217
Pukenui 85

Pullen, Bill 55, 95, 97, 108, 125–26, 131, 132–33, 135, 138, 148, 162–63, 185, 188, 191, 193, 204–205, 208, 216–17, 218–19, 230, 231, 236, **insert 2 pp. 1, 3, insert 3 p. 1, insert 4 pp. 1, 2**; biography 242
Pullen, Gill 108, 135, 148–49, 162–63, **204**, 218–19; biography 242
Pullen, Ian 204
Pullen, Roy 204, **insert 2 p. 3**
Puramahoi 120

Queen Street saleyards 52
Queenstown 16, 28

Raetihi **225**
Raggedy Range 28
rail/railways 10, 42–**43**, 63, 71, 226, 227–28, 233; growth in 10; and work for drovers 43
Rakaia River 33
Rangiahua 82
Rangihaeata 89
Rangiriri 144
Rangitata 111
Rangitata River 33
Rangitikei district 64, 112
Rat Point 138
recipes 208–209
Reefton saddle 164
Rees, William Gilbert 16, 28, 29
Rees run, Lake Wakatipu 16, 28–31, **insert 1 p. 1**
refrigeration 39, 44; *see also* frozen meat trade
rehabilitation (Rehab) farms **58**, 225
Reporoa 138
requirements for droving 96, 98–101
Rhodes, George, and brothers 27, 33
Rhodes, Robert 33
Ribbon, Keith 55, 170–71
Richards, Bill 91; biography 247
Richards, Edwin (Ned) 87–91, **90**, **insert 2 p. 1**
Richards, Harry 87, 91; biography 248
Richards, John (Jack) 87–91, **90**, **120**, 155, 165
Riddle, Billy 95, 188, 194

Riley, Erle 157–58, 229; biography 248
Rimutaka Range 17
rivers 15; crossing 15, 23, 24, 28–29, 30, 31, 32–33, **65**, 74–75, 76, 128, **150**, 153–59, **154**, **158**, **insert 3 p. 1**
Riwaka 89
Road Services 228
roading/roads 41, 63, **133**, **225**, 227; road verges 55, 57, 81; *see also* droving routes; stock routes
Robinson, William 'Ready Money' 99
Ross 138, 214
Rosser, Teddy 195
Rough Ridge 28
routes, droving 15, 60–91
routines, daily 123–26, 139–40; end-of-day 143
Ruawai 195
Ruddell, Sam 156
runholders 98
rural workers 98–100
rural–urban divide 229–32

sales, stock 6, 44–45, 47–53, 80; and drover work 47–49
saleyards 47–53, **49**, **50**, **52**; *see also* individual entries under location
Satchwell, Dave 95
Satchwell, Mike 95
scab 34–35; Scab Ordinance of 1849 34; Sheep Act 1878 34
Shanks, Guy 195, 231–32; biography 243
sharemilkers' herds 44, 136
sheep, 15; from Australia 22; behaviour 127, 129; crossbreeding 40; danger to 20, 33–35; diseases 34–35; lamb trade 40, 45; lambs 30; lame 24; losses 19–20, 23, 27; nature of 31–2; numbers 22, 39, 45; and rail 42–43; store stock 40
sheep dipping 34
sheep inspectors 34
Shennan, Bob 51–53
Shennan Station 30
shepherd *see* rural worker
ships 65–70; coastal fleets 65–68; 'mosquito fleets' 65–68
Shotover River 31
Siberia Station 55

Signal, Norm 147
Signal, Rex 145
Simpson, George 28
skills, drovers' 122–43
Slippery Face 78–79, 117
smoking 78, 128–29
South Canterbury, drove to **4**, 33
South Westland: 100–101, 158, 213, 214; drove (Okuru–Whataroa) 72–81, 117–18, 158, 203, **insert 1 p. 4**; drove map 73; track 41
Sowerby, Henry 246
Sowerby, Julia **101**, 246
Sowerby, Keith 95, **101**, 103, **115**, **184**, 186, 195–96, **207**, 230
spaniard *see* speargrass
Spanish Main 24
speargrass 16, 24, **insert 1 p. 3**
St James Station 185
Stanley, Clutha **226**
statue of drover **92**
Stevens, Mary 105; biography 243
Stevens, Ray 44, 105, 119, 127, 130, 136, 157, 206–207, 218; biography 243
Stewart, Trevor 236
stock: care of 131–35; counting in 143, **insert 3 p. 1**; counting out 123–24; feeding 55, 57, 81, 125, 131–32; injuries 132–35; poisoning 172–75; preferences of drovers 127, 129–30; training 127–30, 138
stock and station agents/firms 6, 47–53, 80, 235, 244; and droving 47; and trader–dealers 53–55
stock routes 41, 125, 136, 165, 166, 225, 230, 231
stockman 9
stockwhip **134**–35
Stonyhurst Station 23
storytelling 11, 80, 196, 214
Stroud, David 139, 230, 234; biography 243
Sullivan, John 100–101, 172, 187, 214–15, 229; biography 248
Sundgren, Bev and Kath 47, 110, 119; biographies 243; *see also* Kath Burton
Sundgren, Charlie ('Sunshine Charlie') 6, 47, 109–10, 182, 195–96, 210

Sutherland, Alex (Alec) 186; biography 248
Swainson, William, Jr. 17, 19, 20–21, 202
Sweet, E. D. 23
Sweetwater 82, 85, 114
Sydney, Grahame 30

Taieri Plain 51–53
Taieri River 153
Takaka Hill 89
Takapau 146
Taneatua 230
Tapawera 88
Tapune, Bob 142
Taranaki 41, 64, **136**, 215
Tasman district 87–91; drove map 88
Te Hapua Station 85
Te Horo tunnel **137**
Te Kao 169
Te Kauwhata 188
Te Paki Station 81
Te Paki Stream 85
Te Teko 138
Tees (ship) 111
Tekapo sale 111, 226
Tekapo Station 104
Temuka 16; sales 111
terrain 17, 28, 31, 35, 72, 85
Thompson, Alex 72, 88
Thompson, Smokey 95, 236
Thompson Butchery 72, 88
Thoms Landing 81
Thomson, John Turnbull 30
Thomson, Ngaire 107; biography 243
Thomson's Gorge 30
Thomson's run 30
Tiffen, Frederick 22
Tokomaru Bay freezing works 45
Tolaga Bay **150**, 170–71
tools 198–221
Tophouse 27, **213**
Torley, Frank 6–7, 244
Totara Flat 101–102, 128
Totara Flat Hotel 215
tourists 79, 105, 157
towns, problems in 165–72; and rural divide 229–32
trader–dealers 53–55; decline of 226, 232–34; and droving work 53
traffic, dealing with 10, 76, 83, 160–65, **163**, **167**, 193–94, 226, 228–29, 230

Traffords Hill **insert 2 p. 3**
Traves, Bill 109, 111, 196, 216; biography 248
Traves, Harold 95, 109, 110–12, 216
trials of droving 31–33
Tripp, Charles 201, 208
trouble, avoiding 125–26, 140–42, 161–62
trucking, growth in 10, 225–29
trucks **198**, 216–218, **226, 227, 238, insert 2 p. 2, insert 4 pp. 2, 4**
tucker boxes 123, 185, 206, 210
Tukituki River 145, 146
Turakina Station **42**
Turimawiwi River 71, 87, 88, 89, 91
Turner, Robin 129, 188, 209; biography 243
tussocklands 15, 16, 22, 40
tutu poisoning 83, 85, 103, 172–75
Tutukaka 118
twilight years of droving 223–34
Twin Bridges **insert 3 p. 2**

Uawa River **150**
Upper Hutt saleyards 52
Upper Moutere 89
Upper Takaka 89
urban–rural divide 229–32

Vavasour, William 17, 19
vegetation 16, 24, 26–27, 33–34, 83, 85, 103, 172–75; burning/burn-off 27, 30, 35, 258
vehicles, for droving 10, 103, 105, 216–19; see also caravans; trucks

Wade, Keith 124
Waerenga Station 55
Waiatoto 75
Waiatoto River 75
Waiho River 76
Waikato 133, 148
Waioeka Gorge **59**, 126, 132, 148, 173, **insert 2 p. 3**
Waipara 144
Waipawa 145, 146
Waipiro Bay 130
Waipukurau 146

Wairangi Station 233
Wairarapa, 15, 17–22, 163; drove map 18; first drove into 17–22; pastoralists 17; sheep numbers 22
Wairau drove map 18
Wairau Plain 22, 23, 27, 212
Wairau River 153–54
Wairoa 112, 113, 131, 171; saleyards 112
Wairoa River 155
Waitapu 89
Waitara freezing works 165
Waiwakaiho saleyards 203
waterways **65–67**, 68, **69, 70, 74**, 150, 153–60, 181–82, 196, 212; see also beaches; bridges; river crossings
Watson, Jim **54**, 95, 215–16
weather 10, 20, 25, 32, 72, 76, 83, 85, 133–34, 210
Weenink, Albert 95
Weld, Frederick 17–18, 19–20, 21, 23, 27
Wellington 22, **167**
Wells, Bobby 85
Wells, Sid 195
West, Joyce 8
West Coast 56–57, 64, 185, 211, 229; goldrush 56–57
Westhaven (Whanganui) Inlet 41, 89, **160**
Whakamaru **insert 2 p. 1**
Whakatane River **insert 3 p. 1**; bridge 156
Whanganui River **65**
Whangarei 81, 82–84, 231
Wharekaka 17, 21
Whataroa 72, 73, 79, 100; droving to 116–17, 214; sale day 80, 100; saleyards 72, 76, 202
whip crackers 102
whistling (commands) 9, 85, 95, 146, 164, 188, 192
White, George 212
White Cliffs (Taranaki) 136, 137
White sisters 248
Whittaker, Kenny 169
Whittaker, Skip 190
Wild Irishman see matagouri
Wilkinson, Charles 153–54
Wilson, Robert 153
Wilson family (Puramahoi) 120
Win, Llewellyn 95

Win, Reg 95
'windmill corner' 138
women 10; drovers 96, 104–108; land girls 10, 105–106
Woodhouse, A. E. 27
Woodville 147
wool industry 15, 35
World War One 104, 111
World War Two 105, 217, 227
Wright, Stan 181
Wright, Vernon, *Stockman Country* 197
Wright Stephenson Co. 50, 68, 110, 116, 244
Wyatt, William 99–100

Yates, Victor 86–87
Yeoman, Donald 244